MW01713332

# Death By Television

## by Ian Johnston

Pottersfield Press
Lawrencetown Beach
Nova Scotia
Canada

**Canadian Cataloguing in Publication Data**

Johnston, Ian.
Death by television
ISBN 1-895900-21-2
1. Television. 2. Television broadcasting. I. Title.
HE8700.4.J63 1998 384.55 C98-950133-7

Credits:
Cover photo by Lesley Choyce
Edited by Judy Wheatcroft
Typeset by Peggy Amirault

Pottersfield Press gratefully acknowledges the ongoing support of the Nova Scotia Department of Education, Cultural Affairs Divsion, as well as the Canada Council for the Arts. We acknowledge the financial support of the Government of Canada through the Book Publishing Industry Development Program for our publishing activities.

Printed in Canada

THE CANADA COUNCIL FOR THE ARTS SINCE 1957 | LE CONSEIL DES ARTS DU CANADA DEPUIS 1957

Pottersfield Press
Lawrencetown Beach
83 Leslie Road, East Lawrencetown
Nova Scotia, B2Z 1P8
Canada
To order: phone 1-800-NIMBUS9 (1-800-646-2879)

To Mom, Murray, and Chuck Jones

Acknowledgements:

Thanks to The Daily News, Sandy, Marilyn, Marla, Michael Weldon,
Dave, Nancy, Robert, Marek, Cliff, Peter, Sherry, Steve,
and everyone else I can't remember.

"Well Beaver, this may be hard for you to believe,
but life isn't exactly like television."

— Ward Cleaver

# Prologue

**FADE IN:**

**LONG SHOT — DESERT. NIGHT.**

A figure of a man suddenly appears in the moonlight, first small, but growing in size as he moves **TOWARD CAMERA**. He's running, and by all appearances has been for some time. His clothes are covered in dirt. His face reads exhaustion. And desperation.

**CLOSE SHOT** on Ian Johnston's face ... **FREEZE.**

**MUSIC BEGINS ....**

**Narrator**

This is Ian Johnston. Once a critic. Specialty — television. Now a fugitive from the law and those who wish his death — a Death By Television. But Ian Johnston is an innocent man. An innocent man in search of the truth. On the run, Johnston dreams of the day he will meet the shadowy man with one testicle — the man he saw fleeing the newspaper the night his life changed. Every day brings a new town, a new identity. But he must keep running. For Johnston is not alone. A producer named Gerald will not rest till he captures Johnston, and makes him pay for his crimes against TV. But Gerald will have to run fast. For Johnston will stay in a place only as long as it's safe. Then he will move on ... They say there's no rest for the wicked. Nor, sometimes, for the innocent. For rest, in Johnston's case, will most certainly bring Death By Television ...

**MUSIC RISES,** as three jagged words are superimposed on Johnston's face, growing in size until they fill the screen.

**DEATH BY TELEVISION**

**TITLE FADES,** immediately replaced by smaller words ...

THIS WEEK'S EPISODE:
"INDIANS AND ONE-EYED SAILORS"

**HOLD** for the musical climax and ...

**FADE OUT**

# I

I grew up on television. I had no choice. It was there when I arrived.

I was born in 1964, after the demise of *Leave It To Beaver* and John F. Kennedy, but before the debut of *The Man From U.N.C.L.E.*

I grew up in Aylmer, a small tobacco-farming community in southern Ontario. On a map, you can find it about twenty kilometers south of Highway 401, right between Tillsonburg (immortalized in that Stompin' Tom Connors song of the same name) and St. Thomas, the town where Jumbo The Elephant was killed by a train a century ago. There's even a life-size statue of the pachyderm on the road leading into St. Thomas to remind travelers of that circus tragedy.

Aylmer's claim to fame is that it's the birthplace of Aylmer canned vegetables. That's what I tell people, anyway. I'm not sure if that's true or if I made it up. I just say it to differentiate my Aylmer (the canned vegetable one) from that other, bigger Aylmer in Quebec.

The Johnstons were a middle class family. Dad was a small town lawyer who eventually gave up his practice to publish a harness racing magazine. Mom was mom, handling the cooking, cleaning and kid duties, while also working at Dad's magazine.

Clear TV memories don't emerge until 1970. Being near Lake Erie left us within antenna-range of American stations from such exotic border locales as Erie, Pennsylvania and Cleveland, Ohio. At a time when most Canadians had two channels or fewer, we had eleven. And by 1970, we also had two TVs, which really gives a kid bragging rights in the school yard.

The newest addition to the family was a funky Realistic portable with a twenty-inch screen, a snooze timer, and a fake wood cabinet and matching TV stand. The adults got the Realistic, but we kids didn't mind. It meant we got the old TV — a massive wooden floor model with cloth-covered speakers, and five vertically-arranged tuning knobs that seemed to work.

It was a big, clunky behemoth that dominated our rec room. You'd twist the on/off switch, and a full thirty seconds would elapse before it'd spring to life. First, the screen dust would crackle with static, and a small dot of light would then appear in the center. Tiny and faint at first, the dot would swell, exploding suddenly to become a horizontal line that expanded quickly to fill the screen. The picture and sound arrived soon after. When in use, that console generated enough heat to keep our cat Tina cozy and snoozing on top of the set through prime time. Or at least until her paws dangled onto the screen, and we'd rouse her with a well-aimed Superball.

Saturday mornings were my must-see TV time, which isn't exactly strange for a six year old. What was strange was that I was usually out of bed by 5:30 A.M., a full thirty minutes before the channels started their broadcast day. But who could sleep? For one thing, I had a parent-legislated bed time of 8:30 P.M. in those days. So by five, I'd already completed my quality sleep time. Anything more was just decadent. I was also too excited to sleep. The cartoons were coming. Seven solid hours of animated mayhem. Snooze and you lose.

So at 5:30, there I'd sit, cross-legged, three feet from the TV, waiting. My only companion at that hour was that big, glowing Indian head set against the backdrop of those squiggly lines, numbers and geometric shapes. For an outsider, it must have looked a little cultish — like I was some junior zombie undergoing cathode-ray brain-washing. *They're all against you, young man. Embrace the test-pattern Indian. He is your only friend!*

When the clock finally hit six, the TV would go black for an instant, springing to life a moment later with the violent adventures of that mumbling one-eyed mariner with the fetish for anorexic women and wilted green vegetables: "I've had alls I can stands, and I can't stands no more."

*Popeye* was followed by the smart-alecky "boids" *Heckle and Jeckle*. Next up was a solid hour of Bugs Bunny and his entourage, followed by a superhero double-shot of *Aquaman* and *Spiderman*, the singing/dancing antics of the *Banana Splits*, the monkeyfied *Get Smart* rip-off *Lancelot Link*, and the latest adventures of that teenage rock band *Josie And The Pussycats*. (Josie — who looked suspiciously like Betty from the Archie cartoons — would eventually take her band skyward in the 1975 cartoon spin-off *Josie And The Pussycats In Outer Space*. The gigs were better, but the travel was murder.)

I didn't like every show. *Josie* was too girlie, *The Flintstones* were too repetitive, and *Rocket Robin Hood* used leftover footage from *Spiderman*. Or was it the other way around? But bad or good, I stayed glued to the tube for the duration. TV had that power.

My Saturday morning ritual was interrupted only by the arrival of my older brother Dave and sister Nancy. Both had program choices of their own. And, in the case of my brother, a considerable weight advantage. So viewing compromises were made. I'd whine, but as long as it was cartoons, I really didn't care.

Breakfast could also be managed without interruption. I could get to the kitchen and back providing I limited my meal to a bowl of cereal (no problem there), and didn't clean up after myself. Which I rarely did. Like most kids, I put more time and effort into avoiding work than work itself.

The marathon would inevitably wrap up with the one o'clock appearance of *Kukla, Fran and Ollie*, a past-his-prime gnome, a buck-toothed dragon, and a middle-aged school teacher. Kiddie TV headliners for decades, the threesome were reduced to introducing bad foreign kiddie flicks by the time I arrived on the scene. Their show was supposed to be "cultural," which means the same as "boring" to the Saturday-morning set.

Cultural or not, it was at this time I'd get my weekly, motherly lecture about squandering my life immersed in the blue glow. I was wasting precious fresh, outdoor air by remaining cooped up inside. That pesky fresh air. Always getting in the way of a good time. "The boob tube is called that for a reason, young man," Mom would say as she pushed me into the afternoon sun, ignoring my protests and snickering over her use of the word "boob."

* * *

My life hasn't changed much since 1970. I don't get up at 5:30 anymore (that's what VCRs are for) but I still sit real close to the set. When TV is good, you gotta be near the action.

My childhood spent in front of the tube turned out to be career training. I not only watch TV now, I write about it. Networks send me shows in advance, and I interview celebrities. People call me TV Guru, and write to me with tube trivia questions. Then at the end of the week, the newspaper gives me a cheque.

Mom no longer bugs me about my tube watching, though she wishes I would stop using bad words in my TV column. Dad never minded my TV-watching in the first place. Heck, he gave up being a lawyer to become a magazine publisher. So what could he say about my journalism career? Someone pointed out to me once that my columns sound a lot like Dad's old magazine editorials. Replace *The Trouble With Tracy* with the trotters and pacers, take out the "bad language," and there's Dad. Spooky stuff, genetics.

It's been suggested that I lead a pathetic life, doing "nothing" but watching TV and then writing smart-alecky comments about it. Others contend I've got a dream job, doing "nothing" and writing smart-alecky comments about it. I can't assess my life. It's the only one I've got. Still, I've never felt guilty about missing out on anything because of my TV habit. In an odd way, I owe TV.

Everything I know I learned from television. And everything I don't know I learned from TV. That's television's beauty — it shows you everything and it shows you nothing. You get too much information and not enough — all at the same time.

Separating my life from television is nearly impossible. My memory cues aren't birthdays, but TV mega-events. Like the night that Kristin was revealed as the trigger person who shot J.R. Ewing on *Dallas*. That was the day I scored a pathetic fifty-five on a

French comprehension test. I didn't watch *Dallas* that night though. I hate *Dallas*.

The finale of *M*A*S*H*? I had rehearsal for a school play that day, and missed dinner. I picked up a cheese burger and fries from a local diner instead, arriving home just as Hawkeye began his opening "I'm disturbed" monologue.

Stuff like that. It's an *idiot savant* sort of thing. I forget my friends' names sometimes, but I always remember where I was and what I was doing when the TV got interesting.

But I do worry sometimes. Is it my life I'm remembering or is it television? Is it TV or is it me? Is it an incident from my childhood or *The Patty Duke Show*? And if I attempt to separate me from television, don't I risk killing us both?

"They laugh alike / They walk alike / Sometimes they even talk alike / They're cousins / Identical cousins ..."

# 2

***Tonight, Rin Tin Tin battles evil poachers threatening a farmer's livestock, while Lt. Ballinger falls afoul of the mob in another action-packed installment of* M Squad.**
***And on Death By Television, fugitive TV reporter Ian Johnston attempts to restore the reputation of a family of TV outcasts ...***

I was brainwashed into hating *Leave It To Beaver.* Ever since I can remember, the media has been ridiculing this series for its idealized portrait of a 1950s American family. "Today's Society Not Like *Leave It To Beaver*" read the headlines for endless articles on single-parent families, teen violence, and other family-based dysfunction. As the media sees it, *Leave It To Beaver* represents the outmoded and restrictive aspirations of another generation. It's a moldy symbol of a past perfect that few, if any, ever experienced. And I accepted this media view, but I really had no alternative. I didn't see so much as a *Leave It To Beaver* rerun until I was well into my cynical teens.

Up until that point, my *Leave It To Beaver* experience was limited to that opening credit sequence, in which the Cleaver family piled into the family car for a pleasant Sunday drive. Out of the door comes the pipe-smoking, tie-wearing dad, Ward Cleaver, followed by his stay-at-home wife June Cleaver in a sun dress. But where are those rascally boys? Here they are! First, it's precocious, good-looking older son Wally, followed quickly by Theodore, the buck-toothed, gullible moppet known to everyone as Beaver.

TV news people loved to run that credit sequence in negative stories about the current state of the American family. The Cleavers were the unattainable ideal — appearing perfect and perfectly silly in that opener. Of course I hated The Cleavers. That was the idea. But it was all a media lie. When I finally got a good look at *Leave It To Beaver*, I immediately identified with it. This wasn't fantasy. This was real life. My life.

The population of Aylmer had been stuck at five thousand for decades. It never got smaller. It never got bigger. When I was growing up, there was one traffic light, one high school, one community pool and one Kentucky Fried Chicken. Cutting Aylmer down the middle was the meager Catfish Creek, or "crick" as it was usually pronounced. Kids tended to stay clear of the creek, except on those rare occasions when we'd hurl rocks at luckless snapping turtles that had emerged from the greenish-brown depths. Aylmer was a sleepy place where the annual Santa Claus parade or Home and Garden Show would grab the front page headlines in *The Aylmer Express*. Still does, actually.

Yet, even a tranquil hometown like Aylmer can hold terrors for children — real and imagined. I was considered a good kid. I studied a lot, paid attention in class, got good marks, and generally stayed out of trouble. But I wasn't good for goodness sake. I was motivated by fear. Fear of bullies. Fear of punishment by teachers and parents. Fear of being different. And fear of all those other multiple, life-altering embarrassments that are a daily childhood reality.

Most kids are sensitive, but I took it too far, living my prepubescent years in terror of making a mistake. I cried with little prompting, and ran to and from school as a defensive measure. But in trying to avoid trouble, it would invariably find me. I'd get beat up for not being sneaky. And if I did go along with some scheme in an effort to conform, it would inevitably turn bad. Throw rocks at the church windows? I dunno. I guess so. As long as everyone else is doing it.

Which made me a lot like Theodore "Beaver" Cleaver, a polite, scared little kid who couldn't run fast enough from T-R-O-U-B-L-E. After all, the real *Leave It To Beaver* isn't about a perfect family's comedic adventures. It's about surviving amid the dangers and anarchy of childhood. *Everything and everyone is against you. Get used to it, kid. Leave It To Beaver* is about kids screwing each other over. Mostly, it's Beav and Wally getting screwed over by other kids because, well, the Cleaver boys aren't as street-wise as the suburban trash of the fictional town of Mayfield. Being considered a nice, well-mannered kid like Wally or Beav is like wearing a "Kick Me" sign in Mayfield. Nice guys finish last, even if they do learn a valuable lesson along the way.

The Beav is particularly vulnerable. He is a nice, timid kid who just wants to have friends and fun. He's been taught by his parents to trust people, leaving him almost defenseless in the face of the supporting cast.

The chief antagonist — and possibly one of the most evil characters ever created for TV — is Eddie Haskell (Ken Osmond), a skinny, hip-talking teenage creep who delights in tormenting the weak in sneaky ways. Eddie's motto — "Nothing's a dirty trick if it's funny enough."

Eddie is Wally's friend, and has been since the second grade. He's that kid we all knew. You didn't like him. You made fun of him to his face. Yet, you hung around him anyway. Why? Because he was always there, and ready to get in trouble. And sometimes, just sometimes, so are you. You just need a little push.

My Eddie was my pal Jim, an older, morally-bankrupt sort who would shame me into skipping school or some other bad-boy activity. "Come on, Ian," he'd say. "Take the afternoon off. What's gonna happen? Nuthin'. Don't be a wimp." Annoying yet handy guy, that Jim.

For Eddie Haskell, a push is never as good as a shove. Eddie likes to knock preschoolers off see-saws, and laugh while they wallow in the mud. When it comes to adults, Eddie's con is to be overly polite. His catch line "You look very lovely today, Mrs. Cleaver" doesn't fool anyone, let alone Ma Cleaver. Maybe it's not supposed to. Eddie compliments adults with such an ironic tone of voice, it's as if he's making fun of them at the same time. What a louse. As Wally so astutely notes, the reason Eddie is such a creep is, well, because "he works at it."

Danger for The Beav comes in less obviously-threatening forms as well. Best friend Larry Mondello is a rotund, fearful redhead with the problems of the world on his shoulders. Larry is not so much bad as dumb, always convincing Beav to do things that can only have unfortunate consequences.

"Let's smoke a pipe. Let's try out your dad's electric drill. Let's skip school." You get the idea. Still, you can't hate poor chubby Larry. His older brother punches him in the arm, his sister calls him a pig, and his mother is a hen-pecked biddy who keeps Larry in line with warnings of his father's impending return. That absent dad — perpetually in Cincinnati — is an ogre. We never meet him, but it's just as well. For Dad Mondello is keeping score. And when he does finally make it home, every past indiscretion of Larry's will result in some serious off-screen punishment. It's payback time, son. The only thing worse than getting cuffed in the head by Dad? Waiting a week to get cuffed in the head by Dad.

Also making life miserable for Beaver is Whitey Witney, a little blond menace who regards Theodore as the only kid small enough to pick on. In his most famous act of childhood malevolence, Whitey convinces Beav to climb a billboard and dive into a giant fake bowl of steaming soup. As Whitey constantly reminds Beav, they're best friends. And when you're best friends, you dive in soup.

Same goes for Gilbert, another conniving liar vying for Beav's attentions. Gilbert likes to brag how his dad is in the FBI, but it's all a ruse. Gilbert overcompensates because his real dad is a traveling musician, who moves the family from town to town. Gilbert needs to make an impression fast in every new town. So he lies. Just another dysfunctional kid in a community where they're falling out of the trees (and into soup).

When Wally and the Beav aren't being tormented by friends, they're being confronted by the realities of life. Even divorce. In one memorable episode, Beav becomes jealous of the presents a friend receives from his divorced parents — both attempting to buy their boy's affection. The episode ends with Beav's friend being called back home by his mom, who is going through a case of "the weepies."

The Beav also comes face-to-face with alcoholism when he befriends a local handyman. Unfortunately, the handy guy is a reformed drinker who is just using the boy to get at Ward's liquor cabinet. Ya can't trust anyone in Beav's world.

Well, maybe Mom and Dad. But they're definitely not the perfect parents portrayed in our smug media. Kindly Ward Cleaver has his flawed moments. When not off working in "the salt mines," Ward likes to wax nostalgic about growing up on a farm with a father who dispensed tool shed discipline. Ward doesn't hit his kids, though he likes to threaten them with stories of his own father's approach to child-raising. Ward also has a quick temper, heading off to his study when Beav does something wrong. One can't help noticing that Ward's study also contains his aforementioned liquor cabinet.

Actually, Ward reminds me a lot of my Dad. Well-liked and easy going most of the time, but with a volcanic temper that would erupt unexpectedly and dissipate quickly. I picked that up from him somehow. It's that genetics thing again. Hmmm. Maybe that's why Dad always watched *Leave It To Beaver* with me when he got home from the magazine "salt mines." That Ward guy must have made a lot of sense to Pop Johnston.

But though no physical punishment is ever unleashed on *Leave It To Beaver*, Wally and The Beav do fear it. Why? Did Ward hit them once? Or is it just all those Dad-hitting stories coming from friends like Larry Mondello? I guess that's all part of being a kid. Life is a daily psychodrama — full of dangers both real and (mostly) imagined. If you didn't think so much, being a kid would be nifty. As the Beav notes in a philosophical moment: "Boy, little kids are sure stupid. I wonder how they ever get brains enough to grow up with."

# 3

***Can you guess the story behind the headlines? Yes, it's time to test your knowledge of current events with CBC's weekly game show with a difference — Front Page Challenge.***

In the early seventies, my constant hope was for Bobby Orr to be the mystery guest on the long-running *Front Page Challenge*. I don't know why. I was a Leaf fan at the time. So why not Norm Ullman or Dave Keon? Perhaps, Bobby Orr was just the biggest "Canadian news maker" I knew at the time. No matter. If Bobby ever made it on *Front Page Challenge*, I never saw him. But I kept watching through the years, maintaining an affection for this long-running low-tech game show.

Viewed today in reruns, *Front Page Challenge* remains unique, a curious combination of serious news and game show schmaltz. Sort of *Meet The Press* and *What's My Line?* rolled into one. For the most part, this curious mix of TV genres works. There were those notable occasions, however, when it just seemed weird. Perhaps there are some figures from history who shouldn't be seen on game shows. Even good game shows ....

**Affable host Fred Davis**: It's an international story panel. An international story. And we'll start with Gordon Sinclair ....

**Gordon**: Did this story happen on a continent beginning and ending in the letter A?

**Mr. X**: Yes

**Gordon**: Asia.

**X:** Yes.

**Gordon**: Was it east of India?

**X**: No.

**Gordon:** Was it in India?

**X**: No.

**Gordon:** West of India?

**X**: Yes.

**Gordon:** Up in the Persia area?

**X**: No.

**Gordon:** Was it in the Holy Land?

**X**: Yes.

**Gordon**: Does this have to do with any Arab uprisings — something in an Arabian country?

**X**: Which question?

**Gordon:** Let's deal with the Arab uprisings,

**X**: No.

**Gordon**: Was it Saudi Arabia?

**X**: Yes.

**Gordon**: Was it the throning of a new king?

**X**: No.

**Gordon**: Was it within the past three years?

**X**: Yes.

**Gordon**: Was it in 1964?

**X**: Yes.

**Gordon**: Well — there was a new king, one abdicated, there was a plane crash, a lot of diamonds were spread around in the sand, they discovered a little more oil, they bought seventeen more Rolls Royces, and married 406 more wives. Any of those stories?

**X**: No.

**Gordon**: Pass to Betty ....

# 4

***We'll return to Front Page Challenge and the identity of Mr. X after these messages....***

*[Four kids march single file into a suburban home, each carrying an oversized, sheet metal briefcase.]*

**Announcer**: It's the Action Kids! What in the world are they up to now?"

*(The A-Kids plop the suitcases on a big kitchen table.)*

**Kid #1**: Let's get down to business.

**Announcer**: What business?

**Kid #1**: Fun business!

*[The Action Kids open their four briefcases, lying them flat on the table. One forms a fort. The others a castle, a military base, and a space center. All come complete with working weaponry, plastic action-figures and vehicles.]*

**Announcer**: It's the Carry-All Action Play Sets from Marx! Look! This one opens up to the wild, action world of Fort Apache!

I'd been watching this commercial for weeks, wishing, hoping-against-hope that the Action Carry-All would show up under the family Christmas tree. A sheet metal suitcase that turns into a fort? It doesn't get much better than that for a six-year-old.

Even at that young age, I was a master at guessing Christmas gifts. I'd dedicate hours to cross-referencing the presents under the tree with the photos in the Eaton's Christmas catalogue. More often than not, I could figure out what I was getting weeks in advance of the opening date. And if the shape and weight and texture didn't provide enough clues, I'd resort to more straight forward espionage. A small hole in the corner of the wrapping might reveal the mystery. Add a piece of tape to the hole later, and no one was the wiser.

The Action Carry-All was simple enough to spot. That square, metallic-sounding box under the tree didn't stump me for more than ten seconds. I knew immediately I'd scored big. And so that Christmas morning, I became the proud, God-like overlord of Fort Apache, with its double wooden gates, detachable towers, horses, "shell-shooting cannons," and soldiers and Indians in every manner of poses. Heck, they even had a plastic dead guy to hang from the lookout tower. It was the coolest. Even if the cannon didn't work all that well.

And best yet, I could just toss all the action pieces into the fort, close it up, and take it on trips. Like into the yard, where Fort Apache transformed into some British settlement in darkest Africa.

But I wasn't the only kid on the block to score an Action Carry-All. My brother Dave got the Cape Kennedy Carry-All, with its miniature rockets, space capsules and working launch pads.

**Announcer**: Just pack the space missile in the launching pad, and ZOOM! Off into space!

Cape Kennedy was niftier weaponry-wise. But it lacked an enemy to wage war with. It was just a bunch of plastic astronauts and technicians. No matter. It wasn't long before Fort Apache had launched an all out attack on Dave's Mercury astronauts in an effort to control the bedroom floor. Not exactly a fair fight, what with all that U.S. government-funded space technology and such. But then again, my plastic horses were twice the size of Dave's plastic John Glenn and Gus Grissom, so I had an advantage of scale.

Recently, I saw that commercial for the Action Carry-Alls again. I was amazed to see how small they actually were. Fort Apache was enormous in my memories. But it was really just briefcase size. I was the small one.

The Carry-Alls have long since disappeared from my possession. But a few of Fort Apache's plastic soldiers got mixed in with my Lego blocks years ago, surviving three decades without a scratch. If you wanted to save anything as a kid, you threw it in with the Lego.

# 5

*When we left Front Page Challenge, Mr. X remained unidentified as Sinclair passed to Betty Kennedy ...*

**Betty Kennedy**: Is this a happy story?

**X**: Yes.

**Betty**: Any record or achievement?

**X**: [hesitates] I don't know.

**Betty**: Any money involved.

**X**: No.

**Betty**: Are you the central figure of the story.

**X**: Yes.

**Betty**: Are you a political figure?

**X**: No.

**Betty**: Are you a military figure?

**X**: [huge smile] No.

[*Audience laughs*]

**Betty**: I don't know why that's getting a chuckle, but it is. Any loss of life in the story — you talk of this being some kind of achievement, but is there any unpleasant aspect to the story?

**X**: Partly.

[*Betty passes, a guest panelist steps in for a few token questions, then main man Pierre Berton takes over.*]

**Pierre**: Is this an archaeological story?

**X**: No.

**Pierre**: Any political aspect to it?

**X**: Partly.

**Pierre:** Is there a discovery of some kind involved?

**X**: Partly.

**Pierre**: Is there a crime of some kind involved?

**X**: Partly.

**Pierre**: Partly a discovery, partly a crime. Saudi Arabia?

[*Bing, bing, bing. Time is up. The Date was Jan. 5, 1963. Mr. X was, in fact, Mr. X. Better known as Malcolm X, who had just returned from Mecca to renounce black racism and his former leader Elijah Mohammed. He would be gunned down a month later. His stumping of the* Front Page Challenge *panel was not cited as a possible motive.*]

# 6

***Next up is CBC's introduction of a new kind of hero — a crusading Toronto coroner in search of the solutions to today's biggest social problems. John Vernon is ....Wojeck!***

*Wojeck* was cool and Canadian, a rare combination on sixties television.

The pock-marked coroner with a sense of justice was an atypical TV hero. Besides his unglamorous looks (star John Vernon will never be mistaken for poster-boy handsome), *Wojeck* was kinda tough to get along with. Sure, he was nice enough in a grumpy sort of way. He just didn't have much of a sense of humor or romance. He'd schedule an inquest on his wedding anniversary, then ignore his wife's protests. If he was on a case, the case was the thing. Unlike most mainstream TV heroes, Wojeck failed. Often. The hour would end, and he'd still be no closer to the truth. Sometimes, Wojeck was barely in his own show.

In one episode that borrowed from *Citizen Kane*, Wojeck monitored an the autopsy on a native-Canadian who had killed himself. Though the audience was treated to flashbacks of the events leading up to the suicide, Wojeck never solved the mystery. To him, it was just another dead man to add to the files.

*Wojeck*'s low budget added a lot to the grim mood. The black and white photography used in the first half of the series and the hand-held camera gave the show an intimate, documentary feel. It was as if you were sneaking a peek at other people's lives, as if you were closer to the action than you wanted to be. It was unsettling and raw.

My favorite moment of Canadian drama is a 1966-67 *Wojeck* episode titled "Pick A Time, Any Time." It combines a dash of sixties hippie cultural, good writing, acting, and psychedelic kitsch.

You'll laugh, you'll cry, you'll stick a needle in your mind's eye ...

## The Opening Freak Out

Director Paul Almond pulls out all the stops with the pre-credit sequence — the bizarre LSD-induced freak out/death scene of rich, bespectacled university student Gordon Plye.

When we first come upon Gordie, he's lying on the university lawn oblivious to everything, staring at the sky. Individual sounds are amplified and distorted. Other students appear and reappear to laugh at poor Plye. The perspective changes suddenly. We see Gordie's point of view as he stares at the sky, his hand, and the school clock. Then the camera turns around for an extreme close-up of Gordie's eye.

The sequence is repeated. Gordie sees himself running in slow motion. There's the sound of a lawn sprinkler, the camera spins, and the clock tower springs to life, striking noon. Then it's back to Gordon Plye's eye as he ceases to be.

Cut to the opening credits. Hope you weren't out getting a sandwich ...

## The Story

Wojeck is called to the university to investigate the death of Plye. As is quickly revealed, the dead student was part of a group that has fallen under the influence of a Timothy Leary-like professor experimenting with LSD.

Wojeck's attempts to get to the bottom of the campus drug problem are handcuffed by the fact that LSD isn't illegal in Canada. The most screwed-up LSD kid (among the living) is Tony Collins (Michael Sarrazin), a suburban, tie-wearing nut job with two girlfriends — one a straight-A student, the other a pregnant, thirty-five year-old jazz singer. Tony wants a life different from his nine-to-five suburban parents. Though exactly what, he's not sure.

Tony alienates everyone with his uncaring attitude and his rambling, drug-fueled pronouncements on life. He also draws the ire of cops by financing jazz chick's failed abortion. With the police closing in, Tony makes one final visit to his guru, pops a handful of LSD-laced sugar cubes and ... well ... I don't want to spoil it just yet.

Beside, that's a lot of plot already. And I didn't even mention the incest angle ...

## Our Man Tony

The script by Philip Hersch is a marvelous piece of writing, never allowing the viewer to choose sides. Oh sure, Tony is quite mad, and quite unaffected by the people he's hurting. Still, there's something in Tony which makes you cheer for him. Reluctantly.

Maybe it's just the casting of Michael Sarrazin, the good-looking, thick-lipped, heavy-lidded anti-star of his day. In tie, mop-top and convertible, Tony is a handsome, charismatic whiz with the ladies. Even when spouting pseudo-intellectual garbage, it's hard not to listen to the guy. He's likable, in an Eddie Haskell, bad boy sort of way. Even Tony's English professor finds the student curious, telling Tony that his essay on Blake has "no logic, yet there's an incredible feeling that it's right."

Wojeck also senses Tony's dual nature. After an encounter with the rambling youth, the coroner comes away confused. "It's like the chicken and the egg. Is the psychosis from taking the LSD, or is he taking LSD because of some previously existing psychotic condition? Or is he psychotic at all?"

Our feelings towards Tony are also influenced by the monstrous adults that occupy this world. Gordie's rich, pushy father is more concerned with Wojeck cutting up his son's body than the reasons behind the boy's death. And Tony's parents are a caricature of the sixties suburban couple. All dolled up in suit and evening dress, they talk about nothing but money and careers.

Daddy Collins has just nailed down the coveted "Reynolds Soup account," and is celebrating by chugging down mixed drinks. Played by *Air Farce* veteran Dave Broadfoot, Mr. Collins is a beastly study in suburban machismo. He's furious when Tony forges his name on a cheque, but softens when he figures out Tony used the money for his girlfriend's abortion.

"Who gives a damn about some square who's too stupid to take care of herself," Dad enthuses.

And let's not forget the "guru" himself, Charles Southern. A part-time psychology professor from southern California, Southern is shown only in shadow or through Venetian blinds as he conducts LSD sessions with his students. But he's no friend of sixties youth. He's a businessman, doing research for his next book, *The Nature of Consciousness-Expanding Drugs*.

"The mind is a bird, let it soar," he burbles. "When you have freed yourself of all your chains, there will be rebirth."

I think I'll wait for the paperback.

Despite these cartoonishly villainous adults as role models, the troubled Tony remains something of a cad in the end. He has no reaction when Sheila says she's pregnant, and he abandons her when the abortion goes awry. He does the same with Liz during her first "freak out." When she cries out for him, he just pops a couple of sugar cubes and splits.

Tony talks a good game, but he's a self-centered coward.

### The Summer of Love — Canadian Style.

American TV always told us the stereotypical sixties actually happened *in* the sixties. But that was not true in Canada as *Wojeck* reminds us. Don't get me wrong. We had the ideals and antiwar attitudes and stuff. We just didn't have that sixties look: the beads, ponchos, torn jeans, long-hair and beards — those accoutrements didn't arrive in force till at least the early seventies. Check it out in an old high school year book or photo album if you don't believe me.

When I worked at a weekly newspaper in southern Ontario, I happened across a photo file of an annual teen dance in Woodstock, Ontario. 1967, 1968, or 1969 — the crowd was filled with short-haired, tie-wearing young folk. Clean cut, bring-home-to-mom-and-dad types. And then boom — 1972! Suddenly, Woodstock the town looked like ... well ... Woodstock the rock concert. It's the Summer of Love in rural Ontario, five years late.

*Wojeck's* "Pick A Time, Any Time" supports my claim. Though made at the height of American hippiedom, in Toronto (a city susceptible to south-of-the-border trends) there's precious little hippieness on display. The hair is a bit longer, in a Fab Four sorta way. But suit jackets and ties are still in the fashion. And the most outward appearance of youthful rebellion seems to be hanging out at a jazz bar, drinking coffee.

Except for Tony, of course. But then again, he's getting his advice from a guy from southern California. Damn Americans corrupting our youth. The only "hippie" anywhere in this *Wojeck* episode is more a fifties beatnik holdover, hanging out in front of the jazz cafe in a flowered shirt and goatee and offering up dialogue like "Hey man" and "That's a groove."

He should get Tony to help him with his psychedelic banter.

Here's a few ideas to get him started ...

## Tonyisms

"There's a moment in everyone's life when you must choose between perpetual life and infinite darkness."

"It's not what's behind the door that's important. It's having the courage to open it" — Tony gives career advice.

English prof: "Your essay has such an intimacy — as if you knew Blake personally. There's such ... oh what's the word, Tony?

Tony: "Oneness."

"Look at this hand. Just like any other hand. It can touch things. It can feel. But have you ever stopped to think what kind of marvelous universe controls this hand? What makes this hand feel, makes it clutch things. Haven't you ever wondered?" — Tony speaks of life.

"Don't call me baby." — Tony talks to his dad.

"He refused to transcend himself" — Tony offers his best guess on the death of Gordon Plye.

### The Great Canadian Kiss

There are two smooches in "Pick A Time, Any Time" that are important in Canadian TV annals. The first is between Tony and Sheila. Sheila is thirty-five and, to tell the truth, hardly a knock-out. She sings at a local café (she performs *Fair Wind Blues* in the episode), and wears gaudy rings on her toes. She's in love with Tony, who seems incapable of love. Still, when they meet, the pair immediately embrace and kiss. For this, director Almond moves the camera in for a hand-held close up, emphasizing the awkward mashing of faces, heavy breathing and slurpy noises.

I see that kiss as an example of Canada's documentary film legacy. Throughout the first part of this century, we made few film dramas because we could not compete with the Americans. Instead, we made documentaries, lots of documentaries. We were the best in the world at that genre. But when we finally got into fiction films en masse in the sixties, they still looked like documentaries. The subject matter was downbeat and gritty, with scenes of homely-looking (dare I say realistic?) people engaged in awkward love-making. The kiss between Tony and Sheila is a perfect example — a clumsy yet intimate suck face. It makes me uncomfortable. Just like any enthusiastic, public display of affection does. It doesn't help that the kiss is topped off with Sheila's announcement that "I'm pregnant." Talk about throwing a bucket of cold reality on the mood.

The second great Canadian kiss comes with Tony's announcement that he has no intention of joining his father's company: "I want something more! There's got to be something more!" he screams at his dad. Mr. Collins responds by diving into his backyard pool. To soothe Tony, mom kisses him on the cheek. But Tony will have none of that. He turns his head, laying a big, passionate, wet frenchie on his mom. Tony's motivation for this kiss?

Attracted to her?

Wants to shock her?

Loony tunes?

All of the above?

Whatever the reason, the kiss shuts his mom up, and she retreats into an alcoholic haze for the rest of the episode.

## And Now The Climax

Remember Wojeck? During all these Tony-related shenanigans, our crusading coroner has been trying to find out what adult is behind all the LSD experimentation, and whether there's any Canadian law being broken.

There isn't.

Wojeck eventually turns up at the Collins' house. Dad is his normal happy, blustering self, denying his son has done anything wrong, or is implicated in Plye's death. Mom is in a post-kiss drunken stupor. And Tony? Well, he's still spouting his philosophy while staring at a patio lantern.

"Each of us is given life and death. It's the fate of some to spend it as dead souls. But it's the destiny of other people to experience psychic rebirth."

Yeah, right. Wojeck doesn't buy that crap either, but what can he do? Tony isn't listening to anyone but his guru, who advises his drug-addled brood to "float down the river of your mind. Float out to the sea of light."

Unfortunately, the LSD that Tony takes hasn't opened up his mind to symbolism. So Tony takes the groovy guru's words literally, and floats out on the nearest sea (Lake Ontario), his arms raised in defiance.

## Epilogue

Man, I still can't believe that hour of TV. There's an anti-drug message in there somewhere. But it also has an anti-establishment, anti-life, anti-parents, and anti-education message. Any institutions left out?

## And Now A Word From The Man Himself ...

When *Wojeck* was in its heyday, actor John Vernon half-jokingly claimed to be thinking of establishing an acting school in Canada.

"Not so much to teach the classics and good modern plays, but how to make something out of nothing — to teach young actors how to perform bad scripts well," he was quoted in a 1966 CBC press release. "For it's a sad fact that to make a decent living in this country, an actor must play in all sorts of vehicles that aren't good or right for him."

Vernon should know. His four-plus decades of acting have included performances on Broadway, countless films from around the world, and TV in Canada and the United States. Some of his acting assignments have been in great projects, some not so great, and some bear titles like *Chained Heat*, *Killer Clowns From Outer Space*, and *Herbie Goes Bananas*.

Vernon has worked, and worked, and worked, and worked. And kept working. Sometimes, it has meant taking on a mountain of crap before a decent role came along. Unfortunately this is not a unique route. This might explain the link between Canadian actors like Christopher Plummer, Lorne Greene, and William Shatner. They always seem so big and theatrical in their performances, no matter whether at Stratford, or in a guest shot on *Kung Fu*. I always chalk it up to the actors' backgrounds on stage and in countless CBC radio and TV dramas during the 1950s.

"Oh god. The CBC was really a wonderful field for anything you can imagine," said Vernon the first time I met him. "Half-hour dramas, hour dramas, ninety minutes. Ibsen, Chekhov, Strinberg, Shaw, Shakespeare, all the moderns. It was really the largest dramatic center in the world, comparable to the BBC and even bigger than *Playhouse 90*-type shows in the U.S."

I can't allow this to become a celebration of classic dramas. To tell the truth, the Vernon movies I remember are the awful, sleazy, bizarre and downright silly ones. Respectable dramas are a dime a dozen. But killer clowns and living, breathing Volkswagens? How many of those do you see?

"The scripts I get are all over the place," said Vernon. "With bad films, you're faced with this dilemma — if the part is good, it's not a disgrace. But you never know until you get into them. If you like the film, you do it and hold your breath. The business is like a toilet seat. It's up and down."

Despite his apparent candor, I was nervous about bringing up some of Vernon's weirder roles. Experience has taught me that actors can be very sensitive. You never know how they might react to a question about a less-than-stellar project from their past. They may get angry, or just clam up for the rest of the interview, offering one or two-word answers.

I didn't expect Vernon to do this, but you never know. And you must remember I am a coward. Vernon's deep rumble of a voice can be quite intimidating, even when he's being friendly. No surprise the guy got cast as the voice of Big Brother in the original *1984* (1956).

So I proceeded cautiously, starting the interview with a couple of casual questions — like the recent airings of reruns of *Wojeck*. But all the time I was itching to bring up that movie where he played a women's prison warden with a hot tub in his office. Or that foreign flick where Vernon sports a golden schlong and a cowboy hat.

Vernon can blame (or thank) the CBC for kick-starting his Hollywood movie career. His early work on *Wojeck* caught the attention of director John Boorman, who hired him for *Point Blank*, a classic tough-guy actioner with Lee Marvin. Vernon played the bad guy who beds Angie Dickinson before Marvin chucks him out a high-rise window.

Vernon said Wojeck was a combination of Toronto coroner Morton Shulman and consumer watch-dog Ralph Nader. "Wojeck is way out there in the way he handles himself. And that's one of the charming things about him. He's always been flawed in terms of gruffness. But his heart is in the right place, and he knows what has to be done.

"*Wojeck* was absolutely the new thing at the time. It was hard-hitting and honest. The topics were simply the topics of the day. We dealt with things like bad senior-citizen homes, abortions, drugs, drunk driving — all issues that are pertinent today. Sure we lost sponsors, but CBC didn't care."

Though Vernon was more than willing to continue with *Wojeck*, the CBC canceled the series after twenty episodes.

"Honestly, I think it (*Wojeck*) just petered out. When you get into a series for TV, as you well know, you've got to have a battery of writers and editors. And I guess maybe they got tired on *Wojeck*."

The makers of the similar American series *Quincy* were later sued by writers who thought they had been ripped off. Vernon was called in to defend the American series.

"I was there to say that long before *Quincy*, we had a coroner series in Canada. I guess a couple of people had submitted an idea for a coroner and thought that *Quincy* had been stolen from them. I don't think they were as successful as Art Buchwald (who successfully sued Eddie Murphy over his *Coming To America* script), but I didn't stick around to find out."

The reference to *Quincy* in this interview was my cue to speed things up and ask about prison hot tubs, and fourteen-karat penises.

Vernon's willingness to take on any role — no matter how wild — really came to the fore in the eighties. In the Linda Blair in-prison epic, *Chained Heat* (1982), Vernon plays a sleazy warden ("Call me Fellini," he demands) who offers special privileges to female prisoners willing to share his office hot tub.

"That office pool," Vernon remembered wistfully. "I've been asked — did I enjoy jumping into a hot tub with two naked women? And I said it was terrible for all of us. All the body make-up and the water. It took two days to shoot. That's a long bath. *Chained Heat*, now that's a cult film. Absolute nonsense. But I tell you, it has a huge following. I'm not proud of it, but I had a good time doing it."

So we got through the hot tub question without a hitch. Now it was on to the penis flick .... *Sweet Movie* is an indescribable, oft-banned feature by Yugoslavian directing-madman Dusan Makavejev (*WR: Mysteries of The Organism*). It's an "art" film — if you interpret incomprehensible plotlines as art. Still, barrier-breaking scenes of vomiting, defecation, sex in sugar, death in chocolate, WW2 footage, and general weirdness means that *Sweet Movie* ain't boring.

Amid this insanity is Vernon as Texas billionaire Mr. Dollars. As the movie opens, Mr. Dollars is marrying the virgin winner of the Miss World 1984 competition (actress Carole Laure). Everything appears relatively normal until the wedding night, when Dollars unveils his golden penis, pissing on his wife while shouting "Yee Haaaww!"

"Dusan is quite a brilliant film-maker but he's so off-the-wall," said Vernon with a laugh. "I had no idea what that (penis) insert was going to be. Dusan called me and said 'John, remember that scene where you wash your wife? Hitler had a fetish. He liked to piss on women.' And I said what have you done? And Dusan said 'I've found you this big Belgian, and when you get up from washing your wife and wave your hat, I cut to the man pissing.'"

And what was Vernon's first reaction to Makavejev's stunt cock insertion?

"I said 'you rotten son of a bitch.' But a nicer guy you won't find."

Aw, memories.

# 7

***Western justice has a new face — a lawman with a unique approach to keeping the peace. Ronald Reagan stars as the legendary Bear River Smith on this week's Death Valley Days ....***

I grew up watching the Watergate hearings. Actually, I grew up cursing the hearings for pre-empting my game shows. Still, even a disinterested kid in Canada could understand the mood sweeping the United States. Making fun of the president became a TV cliché in those days, a predictable punch line on every sitcom and talk show of the era.

Though Ronald Reagan didn't arrive till 1980, the mood hadn't changed much. Besides, Ronnie just looked silly and old. Even at his inauguration when he wasn't old, he looked old. And caked-on cheek rouge and greasy, dyed black hair did little to alter that impression. Neither did falling asleep at public functions. Or joking publicly about launching missiles at Russia. Real funny. He was goofy. He was ancient. And he was our neighbor.

Still, some of the best Reagan moments were saved for his other career — that of a bad Hollywood actor. Even in the rare good movies and TV shows he made, he sucks. At least, I think he sucks. It's hard to tell from a 1990s point-of-view. After two presidential terms, how can you accept Ronnie as anything but Mr. President?

Most Reagan critics pick the flick *Bedtime For Bonzo* as their fave bad Ronnie performance. A monkey, a president. How can you go wrong? Still, I prefer his subtler bad work on television. It's my genre. What can I say?

My personal favorite is a 1965 episode of *Death Valley Days*. The anthology show "where western history comes alive" featured a different story and cast every week. Most of it was moral hokum, wrapped up in sagebrush and petticoats, and served in a low-budget hour. Ronnie hosted the show in the mid-sixties, starring in the occasional episode when the role was just *right* for him. One wonders if Ronnie wasn't thinking about his impending run for the California governorship when he essayed the role of Bear River Smith, the new marshal of a lawless Abilene.

It's certainly not the type of role you'd expect from the Commie-hating, Star Wars-loving hawk. But this was the time of Vietnam, so maybe Ronnie was trying to soften his image. For Bear River Smith is a lawman with a difference — he won't wear a gun. Yep. No guns. Don't use 'em. Don't like 'em. Now are you starting to understand why this is my favorite Reagan performance?

Let's join our story in progress ....

Bear River Smith wanders into town in his sissified city slicker duds, complete with bowler hat, high collar and tie. The locals are not immediately impressed.

**Mr. Prentiss** [a town bigwig]: I hear you threw a lot of lead up in Bear River?

**Smith/Ronnie**: Yup. that's right. I threw up a lot of lead in my past. Some of it hit the wrong fellas. They threw a badge on me because I used to be a cop in New York. Then I forgot what I learned walking the beat in the Bowery. I don't care whether you're called a cop or a marshal, a man who is hired to protect the public has no right being judge and jury.

**Prentiss**: Well, we didn't send all the way to Wyoming to hire a fist fighter.

**Smith**: Well, there's nothing in this [*holds up contract*] that specifies a gunfighter.

And that's just what the old, jowly Smith is — a regular, sure as shootin' fist fighter. Why, he breaks up a saloon brawl between two drunken cowhands by bonking their heads together, Three Stooges like, throwing them out the saloon doors and right onto their horses. What a stunt.

But Smith ain't done — not by a long stretch. The new law soon turns his sights on town psycho Nate Bender, who brags of already killing nine men. But he's no match for Smith who chases the bad guy out of town for insulting saloon-whore Millie.

Gee. All he did was refuse to call her a lady.

Still, the folks of Abilene remain none too pleased with their pistol-packless marshal. Despite his success beating up drunks, public opinion polls show Smith's popularity has sunk lower than a rattlesnake's belly. But you can't blame the town for being a-mite skeptical. It has already lost three lawmen in the past four months. Without a gun, Smith looks like he's a-fixin' to be number four.

So Prentiss tries to impeach him, but Smith won't budge. The proprietor of the local gun and coffin shop ("I get 'em on the way in and out" is his sales pitch) takes a different approach. He offers Smith a free fancy pistol to hide up his sleeve. No dice. Smith don't want nuthin' to do with no secret arms deal neither. Why, the marshal won't even listen to potential First Lady Millie, who thinks he's a dad-burn fool.

**Millie** [admiring his citified duds]: At least you'll look good at your funeral.

That skepticism seems justified when Nate Bender returns to town. That low-down, flea-bitten varmint wants revenge. And if he can impress Jodie Foster as well, even better ...

**Bender**: Hold it right there. I said hold it!

**Smith**: I told you — I don't intend to be your number ten.

**Bender**: And I told ya to crawl and this time you'll do it. That badge don't mean a thing to me [*fires at bottles over Smith's shoulder*].

**Smith**: You're wrong [*begins to move towards Bender*].

**Bender**: The next one goes right through that badge.

**Smith**: I don't think so.

Bender hesitates, allowing Smith to seize the gun and knock the bad guy to the ground. Gee. Maybe Smith's no arms policy was right all along. Nope. While Smith is removing the bullets from Bender's revolver, Bender pulls another gun from his sleeve and plugs our man Smith. Though the secret service men hustle him into a waiting limo, it's too late. Smith has gone to the big round-up in the sky. That skunk Bender doesn't git to enjoy his victory, however. A long-haired stranger in a fringed jacket steps out of the crowd and shoots Bender dead.

**Fringed Man**: You know — maybe Smith was right. Just a little too soon.

**Millie**: Who are you?

**Fringed Man**: Well, I reckon I am your new town marshal. I'm kinda generally known as Wild Bill Hickok.

And so the curtain falls on this sad tale. As luck would have it, Wild Bill was no Smith. Oh, he agreed with his domestic policies, but he also loved guns. So it wasn't long before every bad guy (and a few drunks, free-loaders and immigrants while he was at it) was dead or chased from Abilene. Still, maybe Smith did come along just "a little too soon." Possibly if Smith were around today, his anti-gun policy would be popular. Not in Abilene of course. Probably not in America, either. But I'm sure there's some foreign, bleeding heart burg out there that'd warm to the idea. I hear Britain is pretty into it these days.

# 8

***On tonight's hilarious episode of Get Smart, Agent 99 faces career stagnation head-on ....***

I preferred talking to Agent 99 on the shoe phone. Meeting her face-to-face would have spoiled everything. Afterall, it's been nearly thirty years since *Get Smart* went off the air and, well, that's a long time. Actress Barbara Feldon may still look lovely, but why risk it? I still can't purge my mind of my meeting with Loretta Swit a couple of years back. She was a nice lady but ... I'd rather remember TV starlets as they were — forever beautiful in decade-old reruns.

Besides, what could be better than listening to Feldon's deep, sexy voice coming through the shoe phone lines? Purrr ....

Would you believe Feldon was in Prince Edward Island to shoot a fifty million dollar costume epic starring Robert De Niro, Tom Hanks and Sophia Loren? No? Well, would you believe she was in P.E.I. to star with Ted Danson in a new NBC series set to air in the old *Seinfeld* time slot? No? How about a mid-season replacement series starring Bob Saget and Robert Urich as renegade priests? Still find that hard to swallow? Well, would you believe Feldon was on the island province to shoot a guest shot in an episode of the Canadian family series *Emily of New Moon*?

Yeah? Well, good. That's what she was doing. TV guest shots are pretty well all you see her in nowadays. She prefers part-time TV work. Series TV can be grueling.

"It can be a five-year commitment of thirteen-hour days on a sound stage," said Feldon. "I don't know about you, but I don't find it fulfilling to play the same character for five years."

Feldon's comment aside, the actress did dust off her Agent 99 persona for an unsuccessful 1995 TV revival of *Get Smart*. But for that, she just took the money and ran: "I made an arrangement that I would only have to work one day, every other week. It's a way to have a life." For Feldon, having a life includes living in New York, appearing in the occasional play, or one-shot roles on shows like *Mad About You*, and stints as a commercial voice-over artist. Why not? She can make soup sound sexy.

"Ideally, if I ever to return to series TV, it would be in a recurring role, where you don't have to be in every episode. I've got other things I want to do in my life."

# 9

***Get set for wacky adventures when an uptight dentist inherits a button-pushing monkey with a knack for getting into trouble. Meet the original odd couple of CBS's new hit comedy Me And The Chimp.***

The first series to hook me was *Me And The Chimp*. Few people remember it. It only aired one season (1972) to small audiences and terrible reviews. But I liked it. So there.

The show starred Ted Bessell (Donald from *That Girl!*) as dentist Mike Reynolds, a likable sort with a wife (Anita Gillette), two kids, and a monkey. For reasons only a sitcom writer could fathom, the kids found the chimp hanging out in a park and brought it home as a pet. This being a quarter-century before the movie *Outbreak*, and other disease-carrying monkey dramas, Mom and Dad had no objections to the family's new addition.

Of course, deadly diseases were the least of the Reynolds' problems. Buttons, as he came to be known, liked to push all sorts of .... you guessed it ... buttons. Blenders, door bells, TV. You name it, Buttons pushed it, much to the chagrin of the Reynolds clan.

*Me And The Chimp* was a Brady Bunch-styled family show, with a simian providing the embarrassing, slapstick situations. It wasn't subtle. The most that could be said about it was that at least the monkey didn't talk. Of course, if that monkey had talked, maybe the show would have lasted another season. Still, I liked *Me And The Chimp* when it first aired. So did my sister. But my brother (being four years older, and more TV savvy) was unimpressed, choosing that program to introduce me to the concept of TV criticism. I can still remember the day when he showed me a *London Free Press* story that ridiculed CBS's new mid-season line-up, with particular attention given to *Me And The Chimp*.

I was appalled that someone, somewhere was making fun of my monkey show. But I let that criticism seep into my head. Though I continued to watch *Me And The Chimp*, it was with more discerning eyes. Suddenly, the simian shenanigans weren't half as funny; the slapstick seemed labored and the plots predictable.

I hate it when I'm influenced (even subtlely) by other people's opinions. It sounds strange coming from a TV critic, but I don't respect other people's opinions. And I don't think you should give much weight to mine either. Because I'm just one person and my reviews (while hilarious and insightful) shouldn't be regarded as anything but one guy's opinion. Besides, television critics are freak TV viewers. They watch a lot more TV shows than the average viewer. This makes them more cynical and hypercritical than they really should be.

But I'm not trying to write myself out of job. TV critics can still be fun to read. And once you've grown accustomed to particular TV critics, you can start interpreting their words, and applying them to your nightly TV watching. Here's an example — Ian likes *Me And The Chimp*, but hates *ER*. But you hate *Me And The Chimp*, therefore any advice Ian offers on *ER* should be questioned. That's not to say that you won't hate *ER* as well, but ....

In the case of *Me And The Chimp*, the critic in *The London Free Press* appears to have been correct. I guess it was a lame gimmicky show. I should have saved my TV obsession for a more worthy subject. Or a funnier one.

# 10

***Young Beaver learns a hard lesson about the television industry in this week's episode of Leave It To Beaver.***

[***THE CLEAVER LIVING ROOM***. *Beav and Larry Mondello sit in front of a large, floor-model TV, laughing hysterically.*]

**Larry** [*eating a sandwich*]: Gee, Beaver. Do you think that chimp could drive a car?

**Beav**: Gosh, I don't know Larry. Cars are kinda complicated. The chimp only likes pushing buttons and junk.

**Larry**: But what about the monkeys that went up into outer space? Flying a spaceship's got to be harder than driving a car.

**Beav**: Uh, I dunno. The spaceship monkey has all the men on earth helping him out. So that makes it easy and stuff.

**Larry**: So why don't they build a car a monkey could drive?

**Beav**: 'Cause monkeys don't want to drive because they haven't gotten someplace to go. [*Larry shrugs as a canned laugh track plays.*]

[*Wally and Eddie enter and look at the TV. Their disgust is obvious.*]

**Eddie**: Howdy, saps. What's this? *Wild Kingdom*? Man, that looks like you, Larry. [*Smacks Larry on the top of his crewcut head.*]

**Wally:** Cut it out Eddie. This is Beav's favorite program. *Me And The Chimp*.

**Eddie**: Ain't that appropriate. Look who's watching it — a chimp and a blimp. [*Laugh track plays as Eddie slaps Larry on the head again*.]

**Beav**: Knock it off, ya creep.

**Larry**: Yeah, knock it off, Eddie.

[*Eddie takes the insult from* Beav, *but smacks Larry on the head again. Larry gets up*.]

**Larry:** I got to go, Beav. I'm, uh, late for supper. Mom's gonna holler at me if I'm late again.

**Beav**: See ya, Larry.

**Eddie**: See ya, blimp. Try not to frighten any babies on the way home. Heheheh. [*turns to Beav*] So why are you watching this junk anyway, squirt? Ain't there something better on? I know at my house, we like to watch Walter Cronkite about now. I'm quite a student of current events.

[*Beav, ignoring Eddie, turns back to TV.*]

**Wally**: Aww, cut it out, Eddie. Why ya always have to give Beaver the business?

**Eddie**: I'm trying to make him wise up. You can't have him going into the world and getting slaughtered.

[*Beav turns and glares at Wally and Eddie. Wally, sensing the tension, motions Eddie out of the room.*]

**Wally**: Let's go to the kitchen. Mom put the old TV in there so The Beav and I wouldn't fight each other.

**Eddie**: Lead on, MacDuff.

**Wally**: Cut it out, Eddie. [*Laugh track.*]

[***THE CLEAVER KITCHEN***. *One week later. Ward enters the kitchen, still in his housecoat. June sits watching a cooking show on the kitchen's TV.*]

**Ward**: Good morning June. I didn't know they had shows like that on at [*looks at watch, surprised*] 7 A.M.

**June**: It's something new called specialty channels. They're adding them every day.

**Ward**: What's that noise in the living room? Is Beav up already?

**June**: Oh, he's been there since at least five. I think that's when the cartoons come on.

**Ward** [*looking at June*]: You seem unhappy. Do you think he's watching too much TV?

**June**: Yes I do. He's a small, sickly boy. He should be outside getting some fresh air.

**Ward**: Do you think I should talk to him?

**June**: Oh I don't think that's a good idea. The Beaver is awfully sensitive. Everything you say to him — well — he remembers.

**Ward**: Are you saying I scare the boy?

**June**: To a six-year-old, a father can be pretty frightening at times. You forget what it's like when your father asks you to sit down for a little "talk."

**Ward**: My father wasn't much for talking — unless you count the trips to the woodshed. [*Laugh track.*]

**June**: Oh Ward.

**Ward** [*smiles*]: I'll let it go for now. You never know. Maybe all this TV-watching will turn out to be good for him. He could become a big-time TV executive some day.

**June:** Or a fat, pale, insane derelict wandering the streets. [*Laugh track.*]

[*A knock at the kitchen door. June answers it. It's Eddie with a newspaper under his arm.*]

**Eddie**: Hello, Mrs. Cleaver.

**June**: Hello, Eddie.

**Eddie**: You've done something new to your hair?

**June**: Not a thing. But thanks anyway Eddie.

**Eddie**: Would young Theodore be around?

**June**: Yes, he would. [*Noticing the paper*] Why Eddie, I never knew you read the morning paper.

**Eddie**: At the Haskell household, we start the day by catching up on world events.

**June**: I see. Beaver's in the living room.

**Eddie**: It's been nice chatting with you, Mrs. Cleaver, Mr. Cleaver. [*Eddie exits.*]

**Ward**: Any kid that polite bears watching. [*Laugh track.*]

[***THE CLEAVER LIVING ROOM***. Beav *sits transfixed in front of the television watching Saturday morning cartoons. Enter Eddie.*]

**Eddie**: Hey runt. No monkeys today?

**Beav** [*coldly*]: Hello, Eddie. Wally's still in bed.

**Eddie**: That gives us time to get acquainted, squirt.

**Beav** [*turns to Eddie, suspicious*]: What's up Eddie?

**Eddie**: Gee. Can't a guy just sit and talk with his best friend's brother?

**Beav**: Sure. Exceptin' when the guy always acts like a creep and punches the best friend's brother in the arm. [*Laugh track.*]

**Eddie** [*ignoring him, pulls out paper*]: Well, not today. I just thought you'd like to read something I spied in the *Mayfield Bugle*.

**Beav**: What would I want....

**Eddie**: It's a review of your favorite show.

**Beav** [*suddenly interested*]: *Me And The Chimp*?

*[Eddie clears throat as if preparing for a speech.]*

**Beav**: Cut it out Eddie.

**Eddie** [*reading from paper*]: "*Me And The Chimp* is a juvenile waste of air time that squanders the questionable talents of Ted Bessell with a contrived premise that has him exchanging one-liners with a simian." [*pause*]. Shall I go on?

[*Beav gets up and peers at the newspaper.*]

**Eddie**: "*Me And The Chimp* makes you pine for the low-brow antics of *Mr. Ed*. At least that horse had personality. *The Chimp* should limp through October, then be canceled. By then, all but six year olds will have tuned out this puerile nonsense."

**Beav**: Wha' duz canceled mean?

**Eddie**: It means croaked. This guy thinks the show stinks and it's gonna be taken off TV.

**Beav**: Aww, you're a crumb, Eddie. [*Turns back to TV.*] Leave me alone.

**Eddie**: Keep watching the tube, Beav. I hear *Magilla Gorilla* is on next. Ha!

[*Beav has had enough. Though he is much smaller than Eddie, The Beav begins swinging wildly at the mouthy teen. A coward at heart, Eddie covers himself up and begins squealing like a girl.*]

**Eddie**: Hey! Cut it out, ya wise guy. I was only giving ya the business.

[*Roused by the noise, June and Ward come into the room and break the boys apart.*]

**June**: Beav! What's going on?

**Ward**: Now Beaver. That's enough. You shouldn't treat guests like that ... even Eddie. [*Laugh track.*]

**Beav** [*sobbing*]: But he ... he ...

**Eddie**: I tried to stop him, Mr. Cleaver. But he was like a madman. Maybe its all the Count Chocula he eats. [*Laugh track.*]

**June**: Oh, Beaver.

**Ward**: Now, I want you to go to your room, Beav. There will be no more TV today or tomorrow.

**Beav**: But my show! It's on tonight. *Me And The Chimp...*

**Ward**: That's enough, Beaver. Go to your room.

**Beav**: But ... but ... [*resigned*] yes sir.

[*Beav skulks off, head bowed.*]

[***WALLY AND BEAV'S BEDROOM***. Beav *enters and jumps in bed, stifling his sobs.*]

**Wally** [*roused from sleep, looks at Beav*]: Gee, Beaver. What time is it?

**Beav**: 8 A.M.

**Wally**: Gosh. I thought I'd missed the whole day. What brings you back to bed? TV busted? Hey, you're not going to slobber and junk, are ya?

**Beav**: No. Hey Wally — why do you hang around a creep like Eddie?

**Wally**: Gosh, a guy's gotta have a best friend, even if he is a creep. [*Laugh track.*] What did Eddie do now?

**Beav**: He showed me a newspaper that said my show was no good and it wasn't going to be on any more.

**Wally**: What? *Me And The Chimp*?

**Beav**: Yeah. So then I socked him, and Mom and Dad saw, and now Dad says I can't watch my show tonight. [*Lays face down on pillow.*]

**Wally**: Gosh. A little kid can get into a lot of trouble before breakfast. Kinda makes me glad I slept in. [*Laugh track.*]

[***CLEAVER LIVING ROOM.*** *One week later.* Beav sits *in front of the TV, June and Wally are perched on the couch behind.*]

**June**: Ward! You better hurry, dear. The show's almost on.

**Ward** [*voice from kitchen*]: What's that?

**June**: *Me And The Chimp*.

**Ward**: Me and the what?

**Wally**: You know, Pop — the show that made Beav sock Eddie. [Laugh track.]

**June**: Oh Wally. Don't bring that up again.

**Wally**: Turn it up Beav. It's starting.

[*Beav reaches for the volume knob…*]

**Announcer's Voice**: The regularly-schedule program, *Me And The Chimp*, has been pre-empted. In its place, we bring you Mutual of Omaha's *Wild Kingdom*.

**Beav**: What? They've canceled my show. Eddie was right. They've canceled my show. [*Runs out of the room.*]

**June**: Oh Beaver. [*Turns to husband as he enters room.*] Ward, do something.

**Ward**: What am I supposed to do? I'm not a network executive. [*Laugh track.*]

**June**: Go talk to him. Make him feel better.

[*Ward hesitates. Still unsure what he is going to do, he follows Beav up to his room.*]

**Wally**: Gee, since the Beav's show isn't on, can I watch *O'Hara*?

**June**: Oh Wally. Can't you show a little sympathy? Your brother's very upset.

**Wally**: Well, gee. I feel sorry for the Beav and stuff. But a guy's gotta take advantage when the opportunity arises.

**June**: [*looking at him suspiciously*]: Where did you learn that phrase? I think you've been hanging around Eddie Haskell a little too much. [*Laugh track.*]

[***BEAV'S BEDROOM.*** *Beav lies face down on bed. Ward enters and sits at foot of bed.*]

**Ward:** Beaver, sit up. I want to have a word with you.

**Beav**: Gee, Dad. I don't feel so good. I don't think I can take being hollered at and junk.

**Ward** [*starts to yell*]: I'm not going to ... [*eases off*]. I'm not mad, Beaver. I just think you may be getting upset over nothing.

**Beav** [*sits up*]: Nothing? It can't get much worser. I missed watching my show last week and then ... and then it's canceled. Just like Eddie says. I won't never see it again.

**Ward**: Well, Eddie doesn't know everything. And just because something is in a newspaper, doesn't mean it's necessarily true.

**Beav**: Don't newspapers give you facts and tell you what's gonna happen?

**Ward**: Well, yes. They do. But newspapers are more than that. They also have things called "columns" in which people like you and me might say what they think.

**Beav**: But a guy like that — when he says a show is going to be canceled, then he should know, right?

**Ward**: Maybe. Or maybe not. It's just one person's opinion. As you get older, you'll discover that adults don't always have the right answers.

**Beav**: Really? Gee ....

**Ward**: Now, I wouldn't be surprised if your show comes back on TV sometime soon.

**Beav**: But what about what the guy on TV said? He said it's not coming on again.

**Ward**: No, what the man on TV said was "pre-empted." That's not the same as canceled.

**Beav**: What's pre-empted, Dad?

**Ward**: It's a TV industry word that could mean a lot of things. Maybe the network made a last-minute change to the schedule. Or perhaps *Me And The Chimp* is just taking a few weeks off, and it'll be back next month.

**Beav**: So maybe *Me And The Chimp* won't be canceled?

**Ward**: Oh, I wouldn't go that far. Pre-empted can also mean canceled ... in some cases.

**Beav**: Oh. So why wouldn't they just say that if that's what they mean?

**Ward**: Grown-ups don't like to tell the truth all the time, especially TV people.

**Beav**: Gee, being a grown-up sure is complicated.

**Ward**: You have no idea. But buck up, Beav. Even if your show never comes back, there'll be another with cute animals before long. Maybe two or three.

**Beav**: Really, Dad?

**Ward**: I can promise you, Beaver.

**Beav**: How did you get so smart about TV and junk, Dad?

**Ward**: I run into a lot of TV people at the track ...uh... I mean the salt mines. Maybe it's time you should call it a night. It's past eight. The sun will be going down soon.

**Beav**: OK, Dad. [*Turns off light*.] Uh, Dad? Do you think someday I could be a TV critic?

**Ward** [*stops at door*]: Why not? Just make sure you marry a nice doctor to support you. Now go to bed.

# 11

***The fugitive reporter confronts the man who taught him the meaning of funny in an animated installment of Death By Television ...***

Hockey. Movies. Bugs Bunny. Hair spray. Chili. For me, no wintry mid-seventies Saturday was complete without one, two or all five of these elements.

Of course, hockey is mandatory for all Canadian kids. And though it occasionally put a crimp in my Saturday morning cartoon ritual, what the heck. I liked hockey. And I could always catch up on the cartoons during the summer rerun season.

Besides, there was always the late afternoon *Bugs Bunny Road Runner Hour*, a sixty-minute compilation of the best cartoons from Warner Brothers' golden period (1940-1960s). The show aired at 5 P.M. on CBC, assuming it wasn't pre-empted by football or downhill racing. I never missed Bugs. I woved that wascally wabbit.

They say smell is a good memory stimulant. That must explain why hair spray reminds me of Bugs Bunny. Mom and Dad would often go out on Saturday nights. So while the kids watched Bugs,

Mom would be in the adjacent bathroom, shellacking her modest beehive into place with an ozone-destroying can of hair spray. That unmistakable chemical smell would sneak under the bathroom door and fill the living room, hanging there for hours.

Mom and Dad's night out also meant there was a good chance we'd get chili. Mom made kickass chili — not too hot, not too bland. One time, she dared to change her chili recipe (adding pineapples!) and there was nearly a family riot. Don't mess with the masterpiece, Mother.

So before they'd go out, Mom would brew up a chili batch, and we'd eat it by the bowlful in front of the tube. Funny though. The smell of chili doesn't remind me of Bugs. Yet hair spray does. Go figure!

Chuck Jones seemed to enjoy my hockey/chili/hair spray/Bugs story. But with him it was kinda hard to tell. Though one of the masterminds behind Warner Brothers' Bugs, Tweety, Sylvester and Roadrunner cartoons, Jones was pretty low key. When he did venture to say something funny, he delivered the joke in a quiet, deadpan style that made you wonder if it was a joke at all. Maybe that's his public persona. Maybe he's a closet wacky guy. Or maybe, just maybe, he's burned out after years spent talking to fans and journalists, each with a hair spray story to tell.

I first spoke with Jones when he was in his mid-senenties and still surprised by the undying interest in his old cartoons. When he started working at Warner Brothers in 1933, cartoons were considered to be disposable, funny little time-wasters that preceded the big budget, star-filled, live-action features. Jones assumed his work would have a short life, then disappear.

"There wasn't any television at the time. Cartoons played with the feature. And then there wasn't any place for them to go. Theoretically the prints would then be destroyed and that would be it," Jones said. Happily, it didn't happen. TV stepped in, and gave the Warner Brothers cartoons a whole new audience — kids. Young and old.

"We never made them for a particular audience. We never knew what picture the cartoon would play with or who would see it. So we just did them for ourselves."

"Ourselves" included such legends as Tex Avery, Friz Freleng, Robert McKimson and Bob Clampett — all with their own particular style and approach to Daffy, Porky and Bugs. Though Jones' cartoon repertoire boasts many gems, a half-dozen rank with the best in animation. They are also instantly recognizable to every kid and ex-kid who ever sat down to watch cartoons.

The 1994 book, *The 50 Greatest Cartoons*, surveyed a thousand animators, film critics and members of animation societies. The results placed four Jones cartoons among the top five of all time: *What's Opera Doc?*, *Duck Dodgers*, *One Froggy Evening* and the surreal *Duck Amuck*, in which Daffy finds himself at the mercy of a mischievous, unseen animator. All told, nine Jones cartoons appeared in the book's top fifty.

Remember the one where Bugs sings while cutting Elmer's hair (*The Rabbit of Seville*)? Or Daffy's adventure in space (*Duck Dodgers In The 24th and a Half Century*)? Or when Daffy and Bugs fight over whether it's Rabbit Season or Duck Season (*Rabbit Fire*)? And what about *What's Opera Doc?* in which Elmer hunts Bugs to the music of Wagner? "Kill The Wabbit! Kill The Wabbit!" Funny stuff.

My favorite Jones entry remains *One Froggy Evening*, in which a man's attempts to exploit a singing-and-dancing frog constantly go awry. "Hello my baby / Hello my honey / Hello my ragtime gal." Jones said the dialogue-free *One Froggy Evening* came "closest to living up to what I hoped it would. I get a lot of satisfaction from it because it was an extremely difficult film. There's quite a bit of conversation implied in it, but all you hear is the frog. So all the dialogue is done in pantomime behind windows and such. This cartoon really helped me understand what my job was as a director. I had to draw the frog as a real frog so that when he started to dance he was funny. He wasn't like Bugs or Daffy. He couldn't be a humanized frog. Otherwise, there wouldn't be any surprise when he jumped up and sang. He had to be a real frog. And audiences always react to the fact he's a real frog that sings and dances."

Jones claimedas the voice for the Merchant Navyas the voice for the Merchant Navy not to have a favorite cartoon: "It's a bit like having children. You can't pick a favorite child. You love them all for different reasons." His "children" are having a long, healthy life. So is Jones, who still can't believe he's made a life out of drawing funny pictures: "We thought it was absurd that anyone would pay us to draw. We were very happy that anyone would give us money to do what we love. But drawing was a disease to us. A pleasant dis-

You're never going to be the best artist or the best animator. And you can't be president if you draw."

Oh really? I'd vote for Chuck. Wouldn't you?

Popular Latin terms (a la Chuck Jones) for Roadrunner:

1) *Accelerati Incredibulis.*
2) *Hotroddicus Supersonicus*
3) *Speedibux Rex*
4) *Velositus Delectibus*
5) *Delcius Delicius*
6) *Birdibus Zippibus*
7) *Digoutius-Unbelievius*
8) *Tidbittius Velocitus*
9) *Disappearialis Quickius*

For Coyote:

1) *Carnivorous Vulgaris*
2) *Eatibus Anythingus*
3) *Famishus Famishus*
4) *Eatibus Almost Anythingus*
5) *Eatius Birdius*
6) *Famishus Vulgaris*
7) *Eatius Slobbius*
8) *Hungrii Flea Bagius*
9) *Overconfidentii Vulgaris*

***

Aylmer's only movie house, The Strand, was located two blocks from my house, making attendance not just convenient, but irresistible. Besides, admission was only fifty cents in the mid-seventies, and a box of popcorn cost a dime. A kid's allowance could go a long way at The Strand.

Inside, it was dingy and damp, with worn-out seats, and the requisite sticky floor. There were large ugly holes in the ceiling where the plaster should have been. A legend surrounded a long-existing stain in the center of the screen. Was it just Coke? Or was it something more sinister. One kid insisted it was the blood from someone who got caught sneaking in the back door without paying.

I suspect that was just an urban legend perpetuated by a financially-strapped manager.

Suffice to say, The Strand wasn't a first-run theater. *Star Wars* opened there a full year after it played in the big cities. Most of the flicks I sat through on Saturdays were old B-movies, worn, second-run prints and leftover foreign crap. None of which mattered. We'd show up every Saturday, no matter what was playing. Ten-year-old copies of Jerry Lewis solo projects? No problem. A dubbed, Czech version of *Snow White*? I'm there. A faded, ragged print of *Mary Poppins*? Save me a spot in line.

*Godzilla* was, of course, a biggie with the matinee crowd, though I preferred *When Dinosaurs Ruled The Earth* and *The Seventh Voyage Of Sinbad.* Love that sword-wielding skeleton! Very cool. I remember being disturbed by Bruce Dern's plight as a lonely deep-space astronaut in *Silent Running*, being confused by Woody Allen's Jewish humor in *Take The Money And Run*, and being baffled by the drug humor and lack of plot in The Monkees' flick *Head.* As for *The Gorgon*, I don't want to talk about it. That snake-haired lady who could turn English dandies to stone gave me nightmares for weeks.

But movies were only part of the matinee experience at The Strand. Besides the features, there was the requisite Bugs or Woody Woodpecker cartoon, and another installment of those 1940s adventure serials in which the G-Men sported big hats and pants to their chests.

The theater manager also got into the spirit of things, offering up draws, skits and games that had us racing through the aisles in pursuit of free passes to next week's double-bill of *Godzilla Versus The Smog Monster* and *The Patsy*. Mr. Mean (a pimply-faced assistant manager dressed in a black cape) would then arrive on cue to halt all the fun. The only way the audience could get rid of Mr. Mean (and start the movie) was to scream real loud.

> *awwww...*
> I said loud!
> *Awwww...*
> You can do better than that!
> *AWWWWW!*
> I can't hear you!

It must have been a precarious existence being a small town theater owner. The Strand changed hands several times in the seventies, closing for good just as I got old enough to date. The last time I was there I watched a Giant Rampaging Animal double-bill in which Charles Bronson took on the *White Buffalo*, and Marjoe Gortner battled giant rats in *Food Of The Gods*. (Is it just me, or are they gunning down real rats on camera in that flick? Not that I like rats, but ...)

The Strand stands empty still. How sad. Those Saturday matinees were a unique and special experience. Had I known how special, maybe I would have appreciated those afternoons more. Or maybe not. All I know is that The Strand has stayed with me. I credit that home-town theater with fostering an early interest in B-movies, particularly horror and fantasy films. I never miss an opportunity to interview stars I watched on the screen at The Strand.

So I jumped at the chance a few years back to attend two conventions inspired by the old B-movie magazine *Famous Monsters* and its founder Forrest J. Ackerman. Held in Washington and Los Angeles, the events fit the bill of what I consider a dream vacation — a hotel with a bed, a TV, clean towels, and entertainment on site. There were lectures, movie-screenings, and rooms full of obscure videos and memorabilia. There were also lots of nice people to chat with, and the occasional oddball to make fun of. All without going outside.

I know. I know. It's easy to ridicule these events as disturbing love-ins for nerds obsessed with sci-fi trivia. And I suppose they are to a large extent. Yet, where else can you listen to Robert Bloch talk about creating *Psycho*, or Richard Matheson chatting about writing for *The Twilight Zone*? These are people you don't usually see, even as a journalist. So go ahead and call me geeky. I don't care ....

Now well into his eighties, Forrest Ackerman was a sci-fi and fantasy fan when there wasn't even a term for the genre. He joined a science fiction society way back in the 1930s that included such members as soon-to-be-famous-writer Ray Bradbury, and soon-to-be-stop-motion-model-expert Ray Harryhausen, the guy who made the sword-wielding skeleton that so thrilled me. Ackerman also collected movie-star stuff decades before acquiring it occurred to anyone else. His Hollywood house is now an unofficial museum that is a tribute to everything weird and wonderful in horror and sci-fi movies.

Begun in 1958, the magazine *Famous Monsters* is a kiddie-oriented smorgasbord of photos from Ackerman's collection, with lots of good-humored articles and bad puns. Boris Karloff, Lon Chaney (Jr. and Sr.), Bela Lugosi, Peter Cushing and Christopher Lee are *FM* fixtures, staring out from the covers in their frightening, black-and-white glory.

Mothers threw out copies of *Famous Monsters* at record rates in the sixties. Today, the mag appears rather quaint — a pleasant reminder of classic (and not-so-classic) B-movies from the past. Still, the self-described Ackermonster is more than a magazine editor and movie fan. He's a Hollywood hanger-on of the highest rank, having met, partied, and cozied up to most of the long-departed stars of yesteryear. He attended Bela Lugosi's funeral, hobnobbed with Boris Karloff, and is credited with coining the word "sci-fi."

And he's more than willing to share his experiences. With his thin mustache and trademark safari suit, Ackerman looks the part of a showman, tosses out anecdotes and puns with little provocation. "Does anyone want to hear about Peter Lorre?" Ackerman said to start a conversation. "And I was once Ed Wood's literary agent, or should I say illiterary agent." Fans adore him, some a little too much: Ackerman and his magazine get credit for everything from successful marriages to saving lives. Gee, and I just thought it was a cool monster mag.

His convention show-stopper was his H.G. Wells impersonation. In 1938, Ackerman saw the famed writer speak during an American tour. The Wells appearance wasn't being taped, so he made a special effort to remember what Wells said and how he said it. So with a tiny, high-pitched voice, and a measured, dignified speech pattern, Wells emerged from the past: "I am going to speak to you for about an hour. Today — east is west and west is east. And they are coming together — with a bang."

Ackerman's bottomless energy undoubtedly attracted many of the stars to the *FM* conventions. Ray Bradbury couldn't say enough about Ackerman, who shared Bradbury's love for dinosaurs and spaceships at a time when such pursuits were considered kooky.

"He gave me my living when no one else would. I'm here for a very great reason — to celebrate Forrest Ackerman," said Bradbury, sounding surprisingly like other emotional conventioneers.

No longer considered kooks, sci-fi writers are celebrated today by more than us geeky fans. Bradbury likes to tell the story of meeting the seventy original Apollo astronauts, fifty-eight of whom thanked him.

"They were all science fiction fans, see? We are the lifters and the pushers and creators of modern society. And what we're doing and what we're saying is more important than all the intellectual nuts in New York City."

Film and TV conventions like to complement their big guests with industry lesser lights, and Famous Monsters was no exception. Beyond Bradbury and Harryhausen, there was *Batman* Adam West, TV horror host Zacherley, and *Star Trek* background-actress Julie Parish. For lovers of black-and-white horror films, there was Carol Borland, who starred opposite Bela Lugosi in *Mark Of The Vampire* (1935) and then had an affair with the much older actor. Like Ackerman, she attended Lugosi's funeral in 1956.

"I can still see him lying there in his coffin," she recalled tearfully, not realizing just how funny that sounds. I mean, who doesn't imagine the Dracula star in a coffin? That was his schtick.

Occasionally, I was disturbed by the conventions' money grab. You'd see actors like Charles Dierkopp (*Police Woman*), Vernon Wells (*Road Warrior*) and wild-haired Al Lewis (It's Grandpa Munster!) sitting at booths, selling autographed photos of themselves. Why? Did they need the cash? The publicity? Both?

The people continued to line up for autographs. For many, it's why they attend conventions. Though seminars with the stars go on all day, they're not always well-attended. But put the same stars at tables signing autographs, and hundreds would assemble to get scribbles on pieces of paper. This struck me as strange — wouldn't you rather talk to your favorite star than get his signature? Are autographs worth all that much that you'd fly a thousand miles just to stand in line for hours?

But then again, FM conventioneers aren't your typical lot. I hung out with two rabid fans of the Japanese Godzilla movies. They had constructed an apartment/shrine to all things green and lizardy, filling it with posters, toys and six-foot high inflatable versions of the famed Japanese monster. As the beer flowed, the pair even offered up a bizarre video featuring obscure versions of old Godzilla flicks, as well as Godzilla home movies and bloopers. And what does a Godzilla blooper look like? The menacing giant lizard approaches the city and trips over his own feet, falling sideways into a skyscraper, and then backwards into a line of apartment buildings. Now that's comedy.

It was best to avoid the self-styled Rocket Boy, a bearded, middle-aged man in a homemade superhero suit that obviously hadn't been washed in months. R-Boy was at the con to sell his own science fiction videos. "I'm the next Ed Wood! Buy my movie," began and ended his pitch. I don't think Ed Wood set out to be Ed Wood. Didn't Wood want to be the next Orson Welles? Or maybe that's just in the *Ed Wood* movie. Reality meets fiction meets reality once again. I heard later that Rocket Boy was ejected from the convention for unspecified crimes, and was last seen chasing a convention volunteer into the parking lot.

Still, I had to admire these sometimes strange folks for keeping the old movies alive. It's individuals like Ackerman who are preserving Hollywood's past, not the studios. For though one convention was held in the shadow of Universal Studios, efforts to organize an in-studio screening of *The Bride of Frankenstein* went nowhere. The studio didn't even send over movie memorabilia or free admission passes for fans. It was apparent that Universal had no time for old monsters like *Frankenstein*, *The Mummy* and *The Phantom of The Opera*. Afterall, the latest episode of *Sea Quest* was behind schedule and over budget.

# 12

***"The program you are about to see is All In The Family It seeks to throw a humorous spotlight on our frailties, prejudices, and concerns. By making them a source of laughter we hope to show — in a mature fashion — just how absurd they are."***

When I was in the throes of *Me And The Chimp* mania, a much better series was garnering attention. *All In The Family* is one of the greatest sitcoms ever. Even referring to it as a sitcom seems like a put-down. As if you're lumping it in with *Three's Company* and *Family Matters*. But to tell the truth, *All In The Family* had all the sitcom clichés — one set, multiple cameras, the studio audience, and that dining table where no one sits on the side closest to the audience. Still, *All In The Family* stood out right from the beginning. It simply looked "different." Shot on video, the show appeared all bright and shiny. Nowadays, that look is common among sitcoms, but back then ....

Beyond appearances, what made *All In The Family* stand out in those early years was its daring premise. The show centers on a middle-aged loading-dock worker who spouts racist epithets at every turn. "Spic," "kike," "black beauty" — the guy is a hate wordsmith. Archie Bunker (Carol O'Connor) also heaps great cruelty on his loving, if somewhat dim, wife Edith (Jean Stapleton), who he refers to as "dingbat." Viewed today, Archie's interaction with Edith qualifies as verbal abuse. Of course, Archie is an equal opportunity insulter. Daughter Gloria (Sally Struthers) doesn't escape his scorn, nor does her liberal college-going hubbie Mike (Rob Reiner), who Archie dubs "Meathead."

Doesn't exactly sound like a comedy the way I describe it, eh? Especially in today's ultra-sensitive times. A friend told me he can no longer watch the show because of Archie's treatment of Edith. Spousal abuse tales are everywhere today, making us all uncomfortable. That unease may explain why *All In The Family* seldom shows up in reruns. Considering its status as one of the most-loved series of all time, few stations choose to air the show today. Heck, *Who's The Boss?* gets re-aired more than *All In The Family*. I guess viewers can't put up with Archie on a daily basis.

This is somewhat understandable. I remember Mom refused to watch *The Honeymooners*, a definite precursor of *All In The Family*. She disliked the show because "everyone just yells." True enough. And you can't get much more politically incorrect than when Ralph (Jackie Gleason) threatens to send his wife "to the moon." Calling Edith Dingbat is tame by comparison.

Still, it's premature to relegate *All In The Family* to the TV-series graveyard, for the comedy remains one of the funniest, most realistic TV shows ever produced. If you're offended by it, you're not paying attention. After rewatching the show's original 1971 pilot, I'm now convinced everyone should just put all that political correctness aside. Chill out and watch *All In The Family* anyway. It still has a lot to offer.

Penned by creator Norman Lear, the first episode introduces us to the main characters. In many ways, it's a typical *AITF* episode — employing a simple situation (an anniversary brunch) to bring the characters together. The comedy flows out of the dialogue that results when bigot Archie and ardent liberal Mike sit at the same dinner table.

Archie debuts as a fully-formed character. He's just returned from church in a snit after the Reverend Felcher delivered a sermon of "socialist propaganda" that blamed Archie for the world's ills. Never willing to let a good argument pass him by, Mike/Meathead wades into Archieland, blaming the older generation for poverty among minority groups. Whoops. Shouldn't have said that.

"If Spics and Spades want their share of the American dream, they have to get out there and hustle for it," suggests Archie.

Unfazed, Mike questions God's existence. Archie labels his son-in-law a "pinko atheist." And when Edith interferes, she's told to "stifle, you silly dingbat." It doesn't end there. Archie uses the argument with Mike to declare that "black beauties aren't lazy. It's just that their systems are geared a little lower." Mike objects to the term "black beauty," though Edith notes that the phrase is "better than when he said 'coon'."

Edith is what makes Archie's ugly rants so palatable. She takes abuse from Archie over and over again, but remains unaffected. And every once in a while, she lets fly with an inadvertent zinger that deflates her puffed-up husband. Or makes him look even worse. And viewers can handle Archie as long as he's portrayed as a fool. So who better to expose his foolishness than an apparent fool like Edith? At one point, Archie criticizes Mike and Gloria for their heavy petting in the living room. Why, in his day, nothing happened until the wedding night. "And not even then," is Edith's innocent retort.

When Mike tries to take Archie down a peg, he's too obvious. He goes straight for the jugular, undercutting his own argument by going too far. He's a campus protester, with lots of strong opinions, but no practical experience. Even if you agree with him, you don't particularly like him.

Much better at neutralizing Archie is Lionel Jefferson (Mike Evans), the young black neighbor. In the first episode, Lionel makes fun of Archie without letting him know it. To do so, he adopts a slow, shuffling "darkie" persona — presumably one that Archie may have seen in old Hollywood movies. Lionel pretends to be everything the racist Archie thinks he is. Playing it slow and stereotypical reaps great comic rewards. After Mike criticizes Archie for his "black beauties" comment, Archie looks to Lionel for support.

"Mr. Bunker? Prejudiced? Against black people? That's the most ridiculous thing I's ever heard," says Lionel. But before Archie can gloat, Lionel seizes on Bunker's use of the Yiddish word "toches." Archie insists he heard it from a Jewish guy at the docks. But Lionel says it'd be alright if Mr. Bunker is Jewish.

Now Archie is trapped. He isn't Jewish. But if he defends himself too vehemently, then he'll look like the racist he is. And if he isn't prejudiced, then being called a Jew wouldn't be a bad thing, right? Marvelous character this Lionel. Too bad he left the show after the 1974-75 season.

I always thought that TV shows would mature over the years, growing smarter and more significant as the seasons went by. But that didn't happen. TV now deals with adult issues like racism and wife-abuse, but only as plot devices that are seen one week, and quickly abandoned. Creator Lear was walking a fine line on *All In The Family*. Bigots and verbally-abusive husbands aren't funny. Yet, by making Archie buffoonish, it works. *All In The Family*'s great innovation was to give us an unrepentant racist as a main character, and make him funny. That still makes a lot of people uncomfortable.

# 13

***The Partridge Family may have to cancel a charity concert after a skunk stows away on their bus. Join Keith, Shirley and the whole Partridge Family for their latest, scent-sational adventure ....***

I was a smart-ass school kid. I admit it.

But I wasn't your stereotypical, mouthy punk at the back of the classroom. My marks were a little too high to be considered a trouble-making slacker. I sat too far forward — fourth row from the back, tops. I was also overly sensitive and puny. Hardly a tough guy. Yet, I spoke too loudly, and made too many jokes during lessons. I was disorganized and untidy, and gave the impression I didn't care. My teachers spotted these negative traits from the get-go. In grade one, I was cast as the smart-alecky Danny (red-haired Danny Bonaduce) in our grade one production of *The Partridge Family* (1970-74).

A classmate had brought in the latest record by the popular TV singing group. And since we were all familiar with the TV show's plots, the teacher decided to take advantage of this. So we put on a play based on the TV show. It sure beat *Peter Rabbit*. We chose the first season episode where the Partridges encounter a skunk, forcing them to bathe in tomato juice before an important concert. As Danny, I was the pint-sized huckster of the family who got all the great comedy one-liners and played bass guitar.

I'm assuming my classmates saw something in me that said "Danny Partridge." More likely I just made enough noise they had to give me the role. Danny was my hero. He mouthed off at adults and got away with it. What's not to like?

Anyway, it beat playing beleaguered middle-aged agent Reuben Kincaid. We didn't have a suit for Reuben, so the class concocted one out of a cardboard box. Holes were cut for the head and arms, and a tie was drawn on the front with magic marker. Voila! Instant power suit. It worked, though wearing a cardboard box in grade one made Reuben a target for every tiny fist within punching distance. Look! You can punch him and it doesn't hurt!

In the end, my portrayal was convincing on a grade one level. I knew some of the lines (having seen the show), but my air guitar work on *I Think I Love You* was spotty.

Actually, Danny Bonaduce wasn't all that convincing either. And his acting career kinda spiraled downward after *The Partridge Family*. By 1979, Bonaduce was reduced to playing the car-thieving buddy of Christopher Knight (it's Peter, the middle boy from *The Brady Bunch*) in an episode of *CHiPs*. Knight got the dialogue; Danny might as well have been a side of flabby beef.

There was no *Partridge Family* sequel for our class. The record was left in the sun during recess. By the time we returned, it looked more like a plastic ashtray from shop class than a long-play. Oh well. The Kincaid suit had been pummeled beyond repair, anyway.

***

You can trace a pattern of smarty-aleckiness through my public school years. My mom saved all my report cards, and it's astonishing how little change there was in my areas of ineptitude....

Grade three: "Errors due to carelessness," "Legible when he tries," "spelling needs to improve," and "Ian has made an extremely good effort to improve his neatness. He is a very enthusiastic and knowledgeable pupil in many areas. Sometimes he tends to get carried away by his enthusiasm and requires a settling influence."

That's a good name for a rock group. Ladies and Gentlemen, would you please welcome RCA recording artists, Settling Influence!!!

Grade four: "Ian is careless at times," "Gets in too big of a hurry," "Ian becomes very flustered and disorganized at times," and "has trouble handling balls." I think that last comment came from my gym teacher. I hope it did.

Grade five: "When Ian takes his time, work can be neatly done."

Grade six: "Although his written expression is frequently impaired by a lack of formal organization, his curiosity, his perception, and his rapid assimilation of new material are generally exceptional."

***

The coolest report card I ever got came from my grade seven home room teacher. He looked the part of your stereotypical science teacher, having blown one hand to smithereens in some childhood rocket fuel experiment (or so the school yard gossip went).

He had a smart-alecky streak in him as well, devising odd punishments for student indiscretions. If you forgot a text book, he'd make you carry the entire contents of your locker with you for the next week. Everything. Gym shorts included. I liked the guy, and showed it in typical junior high fashion. That's to say — I tried to be as much of a pain as I could. For the most part, we got along alright.

Sure, I would get a detention on occasion. But I also hung out after school to talk to the science guy. This curious relationship was perfectly captured in science guy's convoluted commentary on my report card: "Sometimes Ian's self-confidence seems to lead to overly self-important statements. It is good to be good but you become less when you imply to others how good you are — even if you are. As regards results — they are good."

I love that. Probably the neatest thing anyone ever said about me. I'm good, and I know it. And it's good to be good. But only if you're good and you don't act good. Got that? Good ...

# 14

***Ian hides out as a department store Santa in a special Yuletide edition of Death By Television. On the late show, a plague of zombies threatens mankind in the TV premiere of George Romero's classic Dawn Of The Dead. Due to the graphic nature of the program, viewer discretion is advised....***

You can't have Christmas without death. Just look at the movies we consider Christmas TV traditions:

*It's A Wonderful Life* became a Christmas not-to-be-missed because someone forgot to renew its copyright. Suddenly any TV station could air it for free, and any fly-by-night video company could release it. I guess the movie is considered a "Christmas" flick because its partially set during the holiday. It's certainly not the plot — which is anything but uplifting upon close examination. *It's A Wonderful Life* is all about a guy who yearns to be a big shot but ends up being a big fish in the small pond of Bedford Falls. Every effort by George Bailey to improve himself fails. He plans to see the world, but his dad dies, leaving him the business to run. Love then distracts

George into missing a lucrative business opportunity, and marriage and a family traps him into his nine-to-five life.

There's no way out for George Bailey. He's stuck in Bedford Falls. So when his reputation is threatened by a Bailey business scandal, he contemplates suicide. What choice does he have? His life is tied to the town. If he loses people's respect, he's nothing.

OK, sure. It all ends happily. An angel shows George that everyone in Bedford Falls would have had a crappy life if George hadn't been around to wipe their collective chins. Cue the big party. Still, two hours of grim reality precedes the climactic bash. Just because you sing a couple of choruses of *Auld Lange Syne* at the end doesn't make it a holiday flick.

Of course, *It's A Wonderful Life* is really just a rip-off of Dickens' *A Christmas Carol*, which isn't all that happy-go-lucky either. *A Christmas Carol* puts forward the notion that you can be a complete bastard your whole life, but as long as you repent in your last years — on threat of death — you'll be OK. Who wouldn't reform when faced with death? Maybe I'm being morbid, or thinking too much. And don't get me started on the dark, nasty messages lurking just under the surface of *Rudolph, The Red-Nosed Reindeer* and *Frosty The Snowman*. I could go on for hours ....

Considering my feelings about the holiday, I guess it shouldn't be a surprise that one of my fave Christmas movies is an X-rated, flesh-eating one. Ontario had a very tough movie censor board in the seventies and early eighties. Anything slightly violent in nature was chopped to shreds by the board. When videos hit the market, there was no existing provincial law governing them. Suddenly uncut versions of people's favorite movies were arriving at the corner store.

So it was with much anticipation that I, Dave, and my friend Steve stayed up late Christmas Eve to watch George Romero's uncut version of his zombie classic *Dawn Of The Dead*, complete with the previously missing biker/gut-ripping climax. We had to wait till midnight when Mom and Dad packed it in. We certainly weren't going to crank up this bloody question mark of a video with them in the room. Though my brother and I had seen the cut version at a movie theatre, it was so badly butchered that the audience booed the jumpy, nonsensical climax. Thank you, Ontario Censor Board.

The uncut version was, and still is, quite impressive: funny and gross, with a special, indirect Christmas message as well. The plot concerns the earth being taken over by zombies. A band of survivors grabs a helicopter and searches for a new home. They find it in a huge mall, where everything they need is at their finger tips.

But the zombies follow them. The flesh-eaters surround the mall, clawing at the doors in an attempt to get in. "This is an important place in theis lives," deadpans one human survivor. Never has the word "consumerism" seemed so appropriate as in the flesh-eating world of *Dawn Of The Dead*.

After the zombie epic had unspooled, we all went to bed, visions of massive carnage still dancing in our heads. Since then, Christmas shopping at a mall has never been the same. Every shopper looks a little zombie-like at Christmas. T'is the season.

***

I do have some "traditional" Yuletide memories that don't involve death and destruction, unless you count the wrapping paper carnage.

In the Johnston home, the opening of gifts lasted no more than three minutes. The family would assemble in the front room. It would begin without warning. Dad, having stationed himself at the tree, would begin hurling gifts across the floor to the appropriate receiver, who'd immediately begin tearing at the wrapping. And so it would go. We'd just keep opening gifts till no more were left. There was no time for ooohing and ahhing, or ogling other people's gifts. We just kept our heads down and ripped away till there were no more gifts. I don't know why we chose to open gifts that way. It's not like we were in a hurry on Christmas day. But after a few years of Dad's Christmas-gift cannon, it developed into a Johnston tradition. A weird one, but what traditions aren't a little peculiar?

One explanation may be found in Dad's love of the movie *A Christmas Story*, a funny flick about a kid's Yuletide experiences in the 1940s. In that movie, the kids attack their gifts like rabid dogs, nearly knocking over the tree in their gift-opening frenzy. Maybe Dad saw something from his own childhood in *A Christmas Story*. Perhaps Dad's Christmas cannon was a gift-opening approach that went back decades.

All our frenzied ripping and tearing would inevitably drive our cat, Tina, wacko. She'd dive in and out of the torn paper, and leap from the cardboard boxes. She would go nuts. Her pupils would grow large, her claws would come out, and you'd swear she was grinning. Even at twenty, Tina could still go from zero-to-sixty mph in three seconds when surrounded by a floor full of ribbons and bows. One year, someone made the mistake of wrapping up catnip as a Christmas gift for her, sending her into a drug-addled frenzy that lasted till Boxing Day.

As I said, the Christmas morning revelry tended to be short-lived in the Johnston household. A half-hour after gift opening, Mom would demand we clean up the mess and move our gifts into our individual bedrooms. Heart-warming it wasn't. Yet it was a Johnston Yuletide tradition. When you don't have many, you treasure the few you keep.

# 15

***Regularly scheduled programs have been pre-empted so the CBC can bring you this live, special broadcast.***

Award shows always make bad TV, despite the fact that these ubiquitous TV specials do, afterall, honor the best in the entertainment business. With all that talent on stage and behind the scenes, you'd expect a slick, fast-paced bit of watchable TV. Never happens.

Televised award shows always remind me of that annual minor hockey banquet when I'd don my best shirt and itchy pants in some childhood misconception I was about to win an award for my two-goal performance that season. The rubber chicken dinner would lead directly to the rambling speeches, followed by endless trophy handouts, inter-spliced with bad (and badly delivered) jokes, and uncomfortable silences. Even a profanity-laced speech by former Leaf clown Eddie Shack couldn't save those events.

But that's a small town hockey banquet. What excuse do the Oscar, Emmy, Gemini, and Juno Award nights have? They've got money, talent, and big city know-how, but even the biggest TV award shows remain struggles to watch. The Oscars have never lived up to their hype despite seven decades of trying. I still can't shake that awful 1988 memory of Rob Lowe shimmying with a live action Snow White to a reworked version of CCR's "Proud Mary" (Working! Working! Working For Walt Disney!). And that was the opening number. Pee yew ...

Still, I have to admit to a certain cheesy fascination with bad award shows. After all, I do still remember that Rob Lowe incident of ten years ago. Maybe the older bad TV gets, the more memorable it is.

It's certainly true of the 1979 Juno Awards, which I managed to rescue from a Halifax pawn shop a few years back. Back in the late seventies, an electronic store owner with an early Sony Beta machine began taping Canadian TV indiscriminately. Fifteen years later, he got tired of the tapes, and decided to sell them off. Enter Ian Johnston, a wad of cash, and a healthy interest in discontinued video formats.

Amid the curious array of Beta movies and TV shows from the polyester decade was a tape whose label indicated it was the ACTRA Awards. It wasn't. It was really the 1979 Juno Awards, presented live by a very brave but ill-equipped CBC on March 21, 1979.

The rest, shall we say, is music history (of a sort):

## The Setting

Toronto's cozy O'Keefe Center, all flashing lights and glitter. The podium is an oversized swoosh strip, spelling out the words The Juno Awards in that old computerized lettering once used on Blue Jay jerseys. The awards themselves are oversized, clear plastic, pointy triangles. Two feet in height, they appear capable of impaling any winner who stumbles.

## The Crowd

A tuxedoed and wrap-around disco-skirted group. Bell bottoms abound, as do ruffled shirts undone to the navel — no matter how old the wearer. "Music Box Dancer" guy Frank Mills sports an impressive crop of chest hair. The seating hierarchy seems to be middle-aged, chain-smoking record execs at the front tables, and rock stars at the back and in the dark almost beyond the reach of the CBC cameras (though they do try).

The one exception is the center, front-row table, occupied by Prime Minister Pierre Trudeau (about to go down to defeat at the hands of Joe Clark), date/first-lady-of-guitar Liona Boyd, Sometimes-When-We-Touchy-Feely guy Dan Hill, and Patsy Gallant, who'd embarrassed herself the year before by flashing her pantied behind after winning for her song "From New York To L.A."

## The Mood

Very upbeat. 1978 was the year that Canadian acts broke big time in the U.S., with Anne Murray ("You Needed Me"), Nick Gilder ("Hot Child In The City"), and Hill ("Sometimes When We Touch") scoring on the *Billboard* charts. This cross border success story is mentioned so often during the telecast, it starts sounding like an insult to everyone still toiling in Canada.

## The Host

A nervous Burton Cummings. The chart-topper, coming off his hit "Dream of a Child" album, handles the proceedings with a breezy baffledness, spending the first five minutes making jokes about why he shouldn't be hosting. When a gag flops, he just rolls his eyes or looks pleadingly off stage. Cumming's ice-breaking gags, written by Paul Pelove and soon-to-be *SCTV* regular Rick Moranis, emphasize Canada's inferiority complex:

"This year, Canadians did better internationally than ever before. But the one thing that keeps bringing them home? Do you know how hard it is getting a green card?" A cut to Trudeau shows him guffawing knowingly. The politics of the day (and today) also rears its head, as Cummings asserts, "when I think of Quebec, I think of two things, beautiful women and René Lévesque. But we couldn't use Réne tonight. He only knows one tune."

## The Opening Show-Stopper

Disco Queen Claudja Barry. In a song-and-dance routine that must have seemed bizarre even at the time, the one-hit-wonder shimmies her way through "Boogie Oogie Dancing Shoes." Flashpots explode. The TV screen splits in two, then spins clockwise as dancers in velvet short shorts leap in slo-mo. The coup de grace is a TV first — a platform shoes tap dance. Harder than you'd think.

## The Unofficial Theme

Drugs and more drugs. This being the disco and coke-snorting era, the Juno humor takes on a subtle, drug-addled tone. In summing up 1978, Cummings calls it the "year of the soft, romantic ballad," "the year of the picture disc," and "the year Colombia shipped more gold than ever before" (nudge-nudge). It was also "the year that Keith Richards started working on a record in Canada," a not-so-sly reference to his Toronto conviction on heroin possession.

## Technical Innovations

None, but the CBC does try, oh they try. A loud buzzing sound interrupts the Juno action on several occasions. Handheld cameras swing to rock stars in the audience who aren't where they should be. An announcer's plea to "Please Stand By" is interrupted in mid-sentence. And when ninety-pound rock weakling Nick Gilder accepts his award for Promising Male Vocalist, his speech is inexplicably interrupted by a recording of a heated debate in the House of Commons.

This was also the year when music videos first appeared in large numbers. Non-attendees Gino Vanelli and the band Chilliwack are presented through videos, though Chilliwack can't be heard over that darn buzzing.

## Boob Of The Evening Award

Presenters Alex Lifeson (Rush) and Kim Mitchell (Max Webster) appear to be seeing the evening through the bottom of a wine glass. At one point, the pair decide to go off-script, ad-libbing that instrumentalist Liona Boyd could "blow both of us away with one hand," and then laughing hysterically at their Beavis and Butthead double entendre. "You can't say that on TV," quips Mitchell to audience groans. The duo redeem themselves slightly when they refer to Dan Hill's hit "Sometimes When We Touch" as "Sometimes When We Brunch The Eggs Ain't Done Enough." Hill accepts his award without looking at the giggly twosome.

### Honorable Mention For Boob Of The Evening

According to a CBC news story aired later on *The National*, Trudeau's attendance at the Juno Awards is a less than subtle Liberal strategy to make him appear more down-to-earth and accessible. Bad idea. If anything, Trudeau seems pompous as he presents the Juno Hall of Fame award to country legend Hank Snow. Pierre's rambling introduction isn't even about Snow's incredible life and career, but the necessity of Canadian content laws for protecting our culture. Ho hum.

"I think it's hard to realize but ought to be remembered in these surroundings that music is basically the most interior of arts," is the less than promising beginning to his five-minute speech. Hank Snow's name finally pops up in the last five seconds, though the Snowman doesn't appear too fazed by being left out of his own award presentation.

"I'm so nervous standing with such a great man," remarks the Brooklyn, Nova Scotia native, prompting Trudeau to clap for himself.

### Big Performer

Quebec singer Ginette Reno, a big-voiced, big-boned singer steals the show with an enthusiastic medley. She's handicapped slightly by her choice of material — which features snatches from such 1978 hits as "Love Is In The Air," "Break It To Them Gently," and "Sometimes When We Touch." Ouch ...

### Where-Are-They-Now Department

Nester Pister, Jerry Doucette, Toulouse, The Raes and Patsy Gallant? Maybe I don't want to know.

### The True State Of The Canadian Music Industry ...

The 1979 Juno Awards telecast was immediately followed by an Air Canada commercial touting its daily, discounted flights to California.

# 16

***It's a special family episode of Death By Television, as fugitive Ian Johnston risks capture when he returns to his old homestead...***

For years, television has been blamed for destroying families. In the days before TV, families spent a lot more quality time together, talking, playing, and loving each other. Everything was just peachy-keen like in *Leave It To Beaver*. Blah, blah, blah.

I always thought watching TV with my family was a bonding ritual. That's why I avoided it. Who wants to bond when the TV is on? What are we gonna do? Talk? And who gets the remote? Dad or Mom probably.

Come to think of it, I also take a lone-wolf approach to movie-going. I tend to go to weekend matinees by myself. The crowds are smaller. And besides, why bring a date? You can't talk to her.

Which is a rambling way of saying my memories of family-oriented TV watching/bonding are limited. Like most Canadians, the Johnstons shared that terrible moment in 1972 when the Canadians were humiliated by the Russians in the first game of that oh-so-famous hockey series. I sat in the comfy chair nearest the TV. I remember distinctly slumping in my chair and rapping my knuckles on the floor when the Russians scored their umpteenth goal. No matter. We won the series. Oh Canada.

*Hockey Night In Canada* was a big event in our week in the seventies. And I wasn't exactly going out on Saturday nights back then. Of course, the Leafs inspired some enthusiasm during this period. The team had won the Stanley Cup not too many years past. Add new players like Daryl Sittler and Lanny McDonald, and Leaf hockey in the seventies got pretty exciting.

My most vivid memory is of Mom sitting in her easy chair, complaining vigorously at the latest Leaf screw-up. Middling Leaf forward Pat Boutette was Mom's public enemy numero uno. Her favorite seventies catch phrase wasn't "Sit On It" or "Dy-No-Mite" but "Oh Boutette."

A missed pass.

*Oh Boutette!*

A missed check.

*Oh Boutette!*

A penalty.

*Oh Boutette!*

I'm pouring a drink in the other room.

*"Ian! Come quick. Boutette just blew it again."*

Man, Pat really pissed her off. Jim McKenny could also get her pulse racing on occasion, but he wasn't in the same hate league as Pat.

***

On those mid-seventies occasions when Mom and Dad were out on weekends, we Johnston kids would often stay up late to watch the *Hoolihan and Big Chuck Show*. Technically, my sister and I were supposed to be in bed. But older brother/sitter Dave would often cut us some slack. As long as we scampered off to bed at the first sight of the family car's headlights in the driveway, no one was the wiser.

*Hoolihan and Big Chuck* was one of those local, late-night horror movie shows that were once so popular. It was hosted by two guys who worked at Cleveland's WJW-TV. In between the awful movies they ran (*Curse Of The Swamp Creature*, *The Monitors*), Big

Chuck and Hoolihan would do low-budget comedy sketches that utilized the WJW-TV staff. The city of Cleveland dominated most of their gags. The opening even featured a crude King Kong attacking the WJW building, located "on the beautiful shores of Lake Erie."

H and Big C sketches tended to be low-brow gags of a hit-to-the-groin variety. They were done with good humor and no budget whatsoever. The show's specialty was slapstick and ethnic humor — Polish jokes mostly. But they got around ethnic slur complaints in a couple of ways. For one thing, Big Chuck was Polish, at least I assume he was (his last name was Schodwoski). As well, no ethnic groups were actually named. They'd title the skits "A Certain Ethnic Plumber," or "A Certain Ethnic Gardener," and cast Big Chuck as a dumb guy with a big, European-style mustache, ugly sweater, and Leo Gorcey hat. Big Chuck's most popular character was the ethnic western hero The Kielbasa Kid, which should indicate the level of humor H & Big C dealt with. Spicy sausage found its way into a lot of Big Chuck and Hoolihan knee slappers.

The hosts would also steal shamelessly. They'd often just lip-sync to a popular comedy record. And there was one, oft-repeated sketch about a fisherman who finds a sandwich on the beach. When he takes a bite from it, he's hooked, and pulled kicking and screaming into the sea. I thought it was an original H and Big C gag. Years later I discovered they'd taken the gag from a classic short film. But that was part of H and Big C's charm. They didn't have any money, but so what? This was late-night TV. As long as the joke was cheap, quick and do-able, they'd do it. And to make sure someone laughed, they'd tack on a trademark loud cackle at the end.

It wasn't the greatest show, I suppose. Sort of like a community station with slightly better production values. But for a ten-year-old kid staying up past his bedtime, it seemed perfect. I'd sit on the arm rest of the living room couch, one eye on Big Chuck and another out the window, watching for tell-tale headlights.

Funny thing. *Hoolihan and Big Chuck* isn't just a nostalgic memory from my seventies past. A version of the show is still on. Now called *Big Chuck and Lil' John* (Hoolihan left, and was replaced by little guy John Renaldi), the show is entering its thirty-second season. Not bad for cornball humor, eh?

* * *

As I said earlier, I tried to avoid family TV-watching like the plague. So beyond *Hockey Night In Canada*, I don't have all that many memories of watching TV with Dad. Heck, thanks to Mom's Pat Boutette obsession, I hardly remember Dad even being there during games. Funny that.

For the most part, Dad and my moments of "bonding" were horse-related. When we weren't at the track (Dad being a hands-on harness racing magazine publisher and all), I helped him at Gala Farms, a stable and farm operation on the outskirts of Aylmer that Dad co-owned.

A few of Dad's horses were stabled there, but mostly the stalls were occupied by the race horses of others. Every night, we'd drive out to feed and water the critters, and I'd climb up in the loft and toss hay and straw down to the floor below, while Merle Haggard and Charlie Rich played on the 8-track player.

I whined and complained most of the time. Like all kids my age, I was lazy as hell. What's more, I hated race horses. Nasty, stupid, expensive animals. Definitely not Mr. Ed. They were high strung and big enough to hurt you. I had this recurring nightmare of leading a horse into a stall, only to have it turn on me, kicking me bloody with its hooves as I cowered in the corner. It wasn't a far-fetched idea. These animals were pretty twitchy. You had to treat them like expensive china. Big, stupid expensive china with attitude. Funny though. I now look back fondly at my time spent with Dad at the barn. I must be nuts.

One TV/Dad incident does come to mind, though. *Taxi* star Andy Kaufman was hosting an episode of the *Saturday Night Live* rip-off *Fridays!* This was 1981, and Kaufman had been getting a reputation for his eccentricity. In fact, the last time he was on *Fridays!*, he'd purposely screwed up sketches, ending the show with a brawl between him and crew members. It was all a set-up, of course. But it was played very real. Kaufman loved to keep viewers guessing.

For this show, Kaufman had come up with a new gimmick. After introducing his new fiancée, the gospel singer Kathy Sullivan, Kaufman launched into a long rant about how he'd turned religious. His conversion eventually took over the show, as sketches were interrupted so Kaufman could talk about Jesus.

Dad thought it was real. Afterall, it wasn't conventionally funny. There were no gags. So it must be real. I wasn't so sure, but I argued with him anyway. The debate got pretty heated. Eventually, I just went to bed. Kaufman had succeeded in screwing up our heads. That's probably why we still remember that (purposely) annoying comic — because he made us feel stupid.

# 17

***Award-winning writer Rod Serling helms this week's spooky episode of that perennial audience favorite, Death By Television. In this full moon chiller, our hero hides out as a farm hand while coping with a new obsession ....***

**Serling:** Witness if you will Mr. Ian Johnston, a charter member in the fraternity of dreamers and couch potatoes. But Mr. Johnston finds his flights of TV fancy constantly undone by tongue-cluckers and life's responsibilities. In just a moment, events will conspire to place Johnston in a world without distractions, where his mind can be free to explore his passions — again and again and again. For Mr. Ian Johnston is about to discover .... *The Twilight Zone*.

Have you heard the story of American soldiers held in a POW camp during the Vietnam War? To keep their minds off the hunger, death and pestilence around them, the prisoners recounted episodes of *Star Trek* to each other. They even assigned roles. (Who got to play Sulu?) Then they'd switch the roles, and run the same episode.

Though I can't relate to Vietnam, I can relate to the TV fixation. During the summer of 1981, I became fixated on *The Twilight Zone* while coping with mind-numbing boredom. But unlike the POWs, I didn't replay *The Twilight Zone* episodes on purpose. My brain just took me there. Over and over and over.

The one summer job you could count on in Aylmer was harvesting tobacco. Tobacco was the big money-maker for the area. Still is. So the entire community was involved in tobacco on some level. School was even delayed a couple of weeks in September to accommodate the hundreds of students who were still busy in the fields. I took a tobacco job because everyone else did. Besides, the money was good (forty to fifty bucks a day) and you only had to work six or seven weeks to earn more money than you could possibly spend that winter. This sounds good until you start working in tobacco. After five minutes, you wonder why you didn't take that minimum wage job at the 7-11 in June.

Harvesting tobacco is uncomfortable, dirty, sticky, grotesque, repetitive work. I did it for seven years, and I've got nothing good to say about it. So why did I do it for so long? Because I was stupid and lazy. Over the winter, I'd forget how much I hated the job. In the spring, I wouldn't look for a job. And by July, there was only one job I could take — tobacco.

I was a "primer" — a fancy name for the guy who pulls the leaves off the plants. Each farm generally had five primers. We all worked together on what was uncreatively called a priming machine, a noisy vehicle consisting of two giant wheels and metal piping supporting an engine on a platform seven feet above the field. It was a clunky, ugly rig that jerked and dragged us down the tobacco rows, belching blue engine smoke. Its funny shape (it looked like scaffolding with wheels) allowed it to move through the tobacco field, down five rows at a time, without touching a leaf.

Imagine that you were one of the primers sitting on a metal seat at the back, positioned atop a tiny wheel that carried you down the center of the row. Directly in front was a basket — a walled, thin cubicle open at the front. Burlap "baggies" were attached to the top of the basket's wall, forming a U shape.

Your task was to pull two leaves off the bottom of the plants on the right, as they careened by you. Two leaves. Two leaves. Two.

Two. Two. Two. Two. Two. When your hands were full, you'd deposit your leaves, stems out, into your burlap U.

It was weird. As you lurched forward, you'd be cocooned in the middle of a row of tall green tobacco. You were isolated, unable to talk to even the guy beside you. The only way you could be heard was to yell. And yelling was only recommended when your basket was full. Then and only then you'd yell at the guy on the clutch the universally accepted stopping word, "Whoa!" When the priming machine halted, you'd use the momentum of the sudden stop to propel you to your feet. It was then a matter of tying the top of the full baggie, tossing it deep into the back of the basket, sitting down, securing the next baggie, and waiting till the machine lurched forward again. Two. Two. Two. Two. It was mind-numbing, exhausting work. But you couldn't slack off. If you didn't get your leaves the first time around, you knew, just knew, you'd pay for it in September. For when autumn came, the best primers had only three or four leaves remaining on their plants. Slackers might have five to ten. So it was much better to over-achieve in July when you were fresher than in those cold days of September.

Little things made harvesting tobacco even more annoying. Like the morning dew, which drenched you "through to your underwear" as Stompin' Tom Connors sings in his tobacco ode *Tillsonburg*. But it wasn't just dew that got you. It was tobacco juice mixed with dew. When it hit your eye, it stung like hell. But you couldn't rub it because your hands were covered in dirt and tobacco tar. There were also green, bloated, hairless caterpillars with white eggs clinging to the leaves. You could avoid them if you spotted them first; usually you weren't tipped off to their presence until you felt them squirming between your fingers.

But there was no stopping for sore eyes or squashed bugs. You were paid by the kiln. When that big barn was full of leaves, your day was done. There were no exceptions. You could be done by noon, or sundown. Same money either way.

Lots of primers didn't make it past week one. I was fired from my first farm for sheer incompetence. My fingers were too small to grip the leaves, and my baggies were always untidy (a recurring theme — check my report cards). Still, this was southern Ontario. I got another primer job in two days. Two. Two. Two.

Television was my only recreational activity during harvest. I'd work all day, shower, then collapse on the couch in front of the TV till it was bedtime. I watched everything in prime time, but it was *The Twilight Zone* that repeated in my head in the tobacco field. I'd never seen the series till that summer, when an Erie, Pennsylvania station began airing nightly reruns. It was an eye-opening experience. Every evening, creator Rod Serling offered up these creepy, twenty-two-minute morality plays with a sci-fi and/or horror spin. They were very over-the-top, with simple, gutsy dialogue, hyper acting, and unsettling conclusions. Add the presence of the chain-smoking host Serling and his muscular, idiosyncratic introductions, and I was hooked.

This was a TV world that dared to, well, blow up the world. It was a place where gremlins crawled on airplane wings, where aliens made suburbanites kill each other simply by shutting off the electricity. In *The Twilight Zone*, you might wake up in the morning to find that everyone has gone, or that no one knows your name.

Of course, *The Twilight Zone* wasn't perfect. Some of its creepy tales were pretty obvious from the start. It didn't take a brainiac to know that an alien book titled "To Serve Man" was actually a cookbook. Or that Death would come to an old woman's door in the form of a young, good-looking Robert Redford. Plot subtlety was not a *Twilight Zone* strength.

For me, Rod Serling scored more than he missed. Even in the early eighties, few other shows dared to offer impossible plot predicaments without explanation. Shit happened in *The Twilight Zone*. That's why it was so therapeutic. And when you're exhausted and couch-bound, a half hour of death, muscular dialogue, ironic twists, and happening shit can be pretty attractive.

My favorite episode was *It's A Good Life*, a nasty little fantasy about a little boy named Anthony (*Lost In Space*'s Billy Mumy) who controls a whole town with his mind. Unfortunately, the boy is only six, so his control makes no sense. He kills without warning, turning people into monsters, or setting them afire. To finish off his victims, he wishes these walking horrors into "the cornfield." Where was the cornfield? No one really knew. People just didn't come back. It was like hell with a rural angle. To cope with little Anthony, the townsfolk adopt a weird approach, saying nothing but happy things to the boy, no matter his crime: "It's good what you did to that dog, Anthony. Real good. And tomorrow's going to be a great day."

It's an episode without hope, and without hope of a happy ending. It's also one of the few stories to tap into that dangerous insanity of being a kid. Plus, it had that "cornfield" angle which, to me, seemed totally appropriate at the time. If the cornfield is hell, then the tobacco field is, uh, purgatory? Makes sense to me.

I'd watch all this *Twilight Zone* strangeness every night, and rise at 5 A.M. with Serling's creation still swimming around in my skull. And it would stay there all day. My brain was like a stuck record, repeating the same disconnected lines of dialogue over and over again as the plants rolled by.

When just thinking about last night's episode wasn't enough, I'd concoct alternative episodes, and obsess on plot contrivances. I'd change the characters, and I'd alter the plot lines. I'd create extra scenes, invent ways that the dead or defeated characters could extricate themselves from their predicaments. I even gave a few of my revamped *TZ* scripts a more personal feel ....

**Fade in**....

[*A priming machine emerges at the end of the row bearing only four primers. The machine stops and the men begin loading their baggies on the flatbed.*]

**Ian:** [*noticing something amiss*] Hey! Where's Billy? We must have left him behind. [*Peers down row, seeing nothing.*] Anybody seen Billy?

**Abe**: [*walking up to Ian, scratching head*] Billy? Who's Billy?

**Ian:** [*slightly annoyed*] Stop kidding around, man. You know, Billy Harrington. Our buddy. My buddy. Billy. Quit kidding, man. Did he leave early or something? Was he sick? What's going on?

**Henry:** [*approaching arguing pair*] What's all this nonsense? There's just us four. Who's this Billy guy? You been inhaling too much pesticide, man.

**Ian:** [*now angry*] Who is Billy? Has everyone gone nuts? He's been working with us all summer. You gave him a ride to work this morning. What happened to him? This is no joke. With only four primers we won't fill the kiln before sundown. We've got to find Billy! Billy! Billy!

[*As the other men return to their work, Ian's yelling fades. Suddenly, Rod Serling emerges from the tobacco plants, wearing a suit and holding a cigarette in one hand.*]

**Serling**: Once upon a time, there was a young man named Billy Harrington. Or was there? Did Billy exist at all? And if he did, what forces in the universe could take him from the memory of man, and place him deep inside .... *The Twilight Zone?*

Tonight's episode of *The Twilight Zone* is brought to you by the friendly folks at du Maurier, proud supporters of tobacco priming in southern Ontario for forty-seven years...

**Fade out...**

Stuff like that. I got pretty buggy in the fields, didn't I? Clearly a tobacco field had as much potential as a Serling cornfield. To me anyway.

And with the kiln full and my brain worn out, the tobacco day would end. Off I'd go home to a shower, three cokes, a big dinner, and more tales of death, destruction, and perversity in the Fifth Dimension. Then I'd awake the next morning and gasp! The same day would repeat itself. Spooky.

Sure, I would occasionally attempt to think of other things beyond *TZ*, but nothing stuck. It was annoying. But you can't always tell your brain what to do when it's allowed to roam. And if you do start talking to your brain, you've probably got deeper problems than a Rod Serling fetish.

Which gives me an idea. Hey Rod! Can you help me end this tale with style?

**Serling:** No comment here. No comment at all. I only wanted to introduce you to one of our very special citizens — Ian the primer, who toils in a tobacco field near a place called Aylmer. And if by some strange chance you should run across him, you had best talk only of old TV shows. Because if you do meet Ian, you can be sure, his mind is stuck in *The Twilight Zone*.

# 18

***Tonight, a punk nightclub is the setting for an unsolved murder. Can Quincy find the real killer in time to save a young girl's life? It's anarchy in L.A. on tonight's Quincy ... And on CHiPs, Ponch goes through Saturday-night fever as he dons the polyester suit to convince punkers to love the night life and celebrate!***

Nothing is less hip than an American show trying to act hip.

Of course, television has always been slow to address trends and fads, and even slower to understand them. And you can forget about TV taking on any trend that's the least bit "controversial." Take punk rock, for example. It had its heyday for a couple of years in the late seventies, and then was gone. Just like disco.

But while almost every show of that era features a reference to disco (including *Scooby Doo*, who had a haunted roller disco episode), punk rock remained almost absent from prime time. Maybe its lack of commerciality (few punk acts scored big on the charts) had something to do with it. Or perhaps its dark, violent undercurrent made TV writers and producers nervous. And the few times punk does show up, it's presented as the embodiment of evil. *Quincy* (1976-83) was a pretty good drama, when it put its mind to it. For its era, the show was cutting-edge, taking on the current issues of the day (chemical dumping, government LSD experiments). Star Jack Klugman was always outspoken about making the show better, and the stories relevant. (You shoulda pushed a little harder on those last few seasons of *The Odd Couple*, Jack. Pee yew.)

But *Quincy* got a bit wacky when it came to punk, exposing it as a dangerous, even murderous form of music. Just like The Twist, I guess. A *Quincy* episode titled "Next Stop Nowhere" (1982) features a murder at a punk club. A young, punk fan named Zack is stabbed in the neck with an ice pick while a band called Mayhem performs its song "Get Up (I Want To See You Choke)." Zack's punk friends Molly and Abby are seen fleeing the scene of the murder.

Abby is a normal-looking cutie who has recently embraced punk, much to the chagrin of her straight mom, who doesn't like Abby's torn t-shirt that reads "DESTROY." But Abby is more a punk faker than a true punk. She's into the costume, not the lifestyle. Her friend Molly sees this right away, advising Abby that she needs some "punking up." Of course, Molly is the real bad girl of this story, what with her shaved head and purple eye make-up. Molly also knows how to manipulate Abby, convincing the naive girl that she was the one who killed Zack.

Meanwhile, Zack's death has created a media furor in Los Angeles. A local talk show even holds a debate between punks (including Abby, Molly, and Mayhem lead singer Fly), and anti-punks like Abby's mom, Quincy and comely psychologist Emily Hanover.

The debate — which closely resembles the format for today's tabloid talk shows — culminates with Hanover describing punk fans as mere "soldiers in an insane war." Abby takes exception to this, summing up her life view with a quote from a Mayhem song: "Ain't no tomorrow, only yesterday's pain." Abby's mom is appalled, but there's no reasoning with her daughter, who moves out of the house and in with Molly.

Of course, when Quincy finds evidence pointing a finger at Abby and Molly, the girls must go into hiding. Molly, always the manipulator, attempts to poison Abby to protect the real murderer. Turns out that it was Molly who killed Zack because ... well ... the punk music told her to! And she was really wasted up at the time.

In the end, Quincy uses his forensic skills to reveal the true murderer. Nice girl Abby is found in time to save her life and her relationship with her mom. The last scene finds Quincy in his favorite bar, dancing with Dr. Hanover to Tommy Dorsey.

"Why would people listen to music that makes them hate, when they can listen to music that makes them love?" queries Quincy as the lights go down on the episode. Gee. Wasn't Tommy Dorsey big during World War II? Not a lot of loving going on then, me thinks.

But that *Quincy* episode was far more clear-headed than a 1982 *CHiPS* episode dubbed "Battle Of The Bands." In it, the motorcycle cops Ponch and Jon battle a punk band called Pain, who commit all sorts of chaos while performing at a Battle Of The Bands gig. Pain steals guitars from a van, pushing it over a cliff. Then Pain throws a stolen guitar from a roof to the street below, causing a multi-car accident.

At the Battle of the Bands, Pain completes its rampage by trashing a bathroom, causing a riot, pulling a knife, and then threatening to blow up a vehicle. Eventually, our hero cops Ponch and Jon arrest Pain for multiple crimes, and Ponch ends the episode by crooning his own rendition of Kool And The Gang's "Celebration." Why would people listen to music that makes them want to hate, when they can listen to Erik Estrada (Ponch) singing off key to a disco standard?

I'm being a little bit flip, of course. It's not hard to figure out why prime-time TV embraced disco and not punk. Disco was hugely successful commercially, affecting not only what people listened to, but how they dressed, and how they lived. There was money in disco. Chain stores selling leisure suits and disco paraphernalia had to advertise on TV, afterall. Heck, you could even build a show around a guy who dances at discos (remember 1979's *Makin' It?*)

As for punk, it was always more famous than successful. Few punk acts topped the charts. And as for fashion, there weren't too many thirty-five year old executives out on the town sporting green hair and safety pins. As well, disco's good time lifestyle was far more understandable to TV watchers than punk's anti-everything message. A punk song can make sense, but a punk lifestyle would be awful hard to maintain. And it certainly doesn't lend itself to a sitcom. Besides, punks don't buy clothes from large chain stores; they steal them.

Of course, the irony is that disco's good time message proved far more deadly to people (via drug overdoses and AIDS) than punk ever was. But at least the disco crowd looked good on the road to nowhere ....

# 19

***Things get wacky on this week's Death by Television, as Johnston serves as the judge at a beauty pageant. Or should that be a judge of a beauty pageant telecast? Take a laugh-filled, revealing trip down the runway with our hero on tonight's Death By Television ...***

I've long had a love/hate relationship with beauty pageants. Over the years, I've condemned them as silly, pathetic sexist parades of desperate women who think the route to career-success comes with a crown, a sash, and a gift certificate from Revlon.

They may be right. I pointed out in a 1991 column that eleven out of twenty-four former Miss Canadas had gone on to work in TV journalism. It's a job not entirely unrelated to the skills fostered at a beauty pageant — good looks, nice hair and figure, ability to look earnest and off-the-cuff even when saying something inane and rehearsed.

That column got one reader quite miffed. In a handwritten letter, an unidentified reader claimed to be a TV producer/former beauty pageant contestant. She took exception to my comments, saying I was just jealous. Not of her pageant experience, but of her TV career. Afterall, she made more money than I do. To emphasize her point, several sentences ended with a HA! She even concluded the letter with a triple shot HA! HA! HA!

Ooooh the pain.

She makes more money than I make? Well, obviously. There's just more money to be made in TV. And who the hell gets into the print business to get rich? Heck, I just want to marry a nice doctor who can support my TV-loving lifestyle. As for being bitter, yes I am. Life made me that way. *Cop Rock* made me that way. But beauty-pageant-contestants-turned-TV-producers? No way.

Heck, I'm only against the *idea* of beauty pageants. As televised events: they're marvelous. The talent portion, the question and answer sessions, the bathing suit competition. Whether a small town Cheese or Furrow Queen contest, or a glitzy televised Miss Universe Pageant, comedy just oozes out of every pore of these dorky flesh parades:

REGULARLY SCHEDULED PROGRAMS HAVE BEEN PRE-EMPTED TONIGHT IN ORDER THAT WE MAY BRING YOU THE FOLLOWING CTV SPECIAL PRESENTATION.

**Announcer**: Live from Toronto, it's the Miss Teen Canada Pageant 1982!

## The Setting

A CTV studio in a wintry Toronto. The stage is a giant staircase. At the left is a glittery podium surrounded by flowers. The requisite runway fronts the stage. Call it Broadway tacky. The studio audience is large, enthusiastic, and bathed in such deep red lighting that they are rendered almost invisible. Their enthusiasm appears to be augmented by canned applause. I doubt people clap and cheer so uniformly.

## The Hosts

Remember that awful Canadian show *Circus*, in which real performers from the Big Top were brought into a less smelly studio to perform? If you remember that, then you remember host Sherisse

Laurence, who tonight has replaced her ringmaster outfit with a long flowing gown, and a big head of hair. Huge hair. Darth Vader helmet size. Probably required a couple of cans of hair spray to maintain.

A former Miss Teen Winnipeg, Sherisse is joined in her pageant hosting duties by a tuxedoed Pierre Lalonde, sort of a Canadian version of Jack Jones — a slick host/singer/gun for hire who specializes in cool Broadway numbers and crooner ditties. Also on hand to add some professional sparkle to the proceedings are the Roland and Romaine dancers, a foursome who appeared on just about every Canadian TV special of the day. The men always seem so excited — flailing and contorting their bodies in exaggerated time to the music. These boys really shimmy and shake.

## The Contestants

"Thirty-two beautiful Miss Teen Canada representatives" from every corner of the country vie for the top prize. Though it's a competition, Lalonde makes sure we all remember that these sixteen and seventeen-year-old girls are all "winners" — each having won some regional competition to qualify for the trip to the big leagues.

"There can only be one Miss Teen Canada, yet here tonight you will see thirty-two of Canada's brightest and most beautiful young ladies — ALL OF THEM WINNERS." Of course, by the end of the evening, one contestant will be *more* a winner than the other thirty-one.

## The Rules

Lalonde tells us that the girls will be judged on "interests, personality, involvement in community activities, academic achievement and character." A new category — fitness — has been added this year. Beauty isn't mentioned as a category till three-quarters of the way through the telecast when Lalonde asserts "judges decisions are based not only on beauty, but poise, personality and intellect as well." Funny that. Beauty isn't worthy of mentioning at the start of the show, yet suddenly it's necessary to play down its importance.

## The Judges

There's the expected head-table gaggle of fashion and make-up experts: a former Miss Teen Canada (now a medical technology instructor), former *Fugitive* chaser Barry Morse, and then-amateur light-middleweight champion Shawn O'Sullivan in a form-fitting tux. Why do they always invite boxers to these beauty pageants? A recipe for disaster.

## The Opening Show Stopper

Sashaying down the center stairway, Pierre Lalonde rips into the show's theme — the cheesy tune "Wonderful Day." It's a toe-tapper, to say the least, with the silver-haired singer wrapping his vocal cords around such memorable lyrics as "When the sun is as big as a yellow balloon / Even the sparrows are singing in tune / The world's in a wonderful way / On a wonderful day like today."

And Pierre is only getting started. During breaks in the opening number, Lalonde introduces the contestants. Out they parade in electric pastel gowns of psychedelic green, orange, red, and blue. Their lips are red and wet, their hair big (though not Sherisse big), with a then-stylish part down the middle. Only Miss Prince Albert, Saskatchewan defies the trend, sporting a short, black Joyce DeWitt 'do.

Sherisse struts to the middle of the contestants and let's fly with "Consider Yourself" from the musical *Oliver*. The girls sway self-consciously to the music. "*Consider yourself part of the furniture...*" The Roland and Romaine Dancers make their presence known by flailing and prancing far in excess of anyone else (and somewhat in excess of the song) while Pierre ("All I Need Is A Girl") and Sherisse ("Luck, Be A Lady Tonight") trade tunes.

Fifteen minutes into the show, the big finish/reprise finally arrives as all on stage raise their arms in the air to shout the virtues of the "wonderful day."

We'll be right back ...

## Insert Ad Here

"At Dominion, you'll love us for more than our meat."

## The Preliminary Awards

Miss Prince Edward Island flexes her intellectual muscles to win an award, The *Encyclopedia Britannica* Scholastic Award. Miss P.E.I. — who accepts the honor to the song "Fame" — got top marks on a quiz on Canada. In keeping with the flag-waving theme, the girls come to the stage sporting red knickers and vests, augmented by puffy white blouses. Oh Canada ....

A little later, with the muzak version of Olivia Newton John's *Physical* as a backdrop, the pageant moves to a videotape visit to Toronto's Fitness Institute, where the girls are put through a series of tests. Sporting green, sleeveless tops and short shorts, the girls lie on the floor while the camera hovers above them like that epic hospital scene from *Gone With the Wind.* Officials move among the horizontal females, testing their resting heart rates. The camera cuts to shots of unidentified arms and thighs being squeezed and prodded by oversized calipers measuring body fat. Kinda creepy.

"Physical / Physical / I want to get physical / Let me hear your body talk / Your body talk."

The teens jump on boxes, do girlie push-ups, and touch their toes. And as the fitness competition is narrowed to three girls (Miss Kitchener/Waterloo, Miss Halifax/Dartmouth, Miss Quebec City), the tests get tougher. Big scary machines measure pulmonary functions, and an electrocardiogram measures ... uh ... electrocardiogram junk. It all ends with a photo-op of the three finalists on matching stationary bikes smiling into the camera.

Back to the studio, and the winner of Miss Fitness is .... Miss Halifax/Dartmouth. However, being fit doesn't get you a cup of coffee in this competition. Neither Miss Fitness, nor Miss Friendship will make the Final Four. Just goes to show you how far a smart, clean-living, healthy girl will go in this crazy world of ours ....

## The First Four Finalists

Though all the girls are winners, the finalists get more goodies. Miss Edmonton, Miss Kitchener/Waterloo, Miss Brantford, and Miss Quebec City are the first named to move on to the second round.

## The Horror

Who doesn't like the question-and-answer portion of a live beauty pageant? No matter how vapid the head-scratcher, something unexpected always happens when contestants are forced to open their mouths and speak in sentences. And screw-ups occur at Miss Teen Canada even when the contestants know the question — the less-than-brain-teasing "Speak on a subject that is important to you."

All four of our first semifinalists sputter when the pressure is on. Like dominos they fall, one after another:

Miss Edmonton (sixteen, plans career in commerce or computer science) is the smoothest, though her speech ("Something that I'm really concerned about is world peace") is standard issue, beauty-pageant fare.

Miss Kitchener/Waterloo (seventeen, career in social work or law) begins promisingly, touting her family as the most important thing in her life. But she goes blank after only two sentences, barely recovering enough to whisper a thank you before sculking off.

A speech about the importance of school by Miss Quebec City (seventeen, psychology) falls apart after only one sentence. And when she attempts to repeat the same speech in French, shc falls apart again.

And then there's the sad case of Miss Brantford (seventeen, math and/or computer science): My individual freedom is very important to me as a young Canadian. And I'm sure it is very important to young Canadians all across ... uh ... Canada. I appreciate the freedom to make my own decisions, and my education, my social activities and my moral standards. This freedom ... this freedom ... umm [*looks at Lalonde*] sorry .... to be ... uh ... able to ... set goals for myself ... and encouragement to ... uh ... [*looks at Lalonde*] sorry."

Lalonde suddenly earns his pageant-hosting bucks by stepping in with a reassuring "You're doing great," even inciting the crowd to clap. But it's to no avail. Miss Brantford is going down. Finally, she just gives up and moves back into the semifinalist line, her hopes of glory gone up in smoke.

## More Finalists

The second batch of semifinalists fare better, apparently learning from the previous contestants' meltdowns. Miss Niagara Falls (sixteen, science), Miss Hamilton (seventeen, early child care), Miss Duncan, British Columbia (seventeen, nursing) and Miss Prince George, B.C. (seventeen, nursing) all deliver relatively flawless odes to education. Miss Niagara Falls sees her future career in pharmacy; Miss Hamilton stutters slightly as she talks of her aspirations in child care; and Miss Duncan talks of working with the handicapped.

But the standout is the dark-haired Miss Prince George, whose speech is well-rehearsed, vapid, and judge-friendly, calling a sound education "the essence to growth." But wait. Miss Prince George isn't done. She gives her speech again in French (very bad French), stealing some of Miss Quebec City's bilingual thunder. Gutsy move.

## The Year That Was

Sporting a full-length wolf sections coat, Miss Teen Canada 1981 fulfills her final duties by showing up in anticipation of handing over her crown. A highlight reel shows her in her home-town of Owen Sound (where you need a wolf sections coat, I suppose), and signing the mayor's VIP registry. Causing one to ponder — just who the hell would be in the Owen Sound VIP registry anyway? Al Capone on a hunting trip? Sir John Franklin in search of the Northwest Passage?

## And Finally, The Four Finalists Are ....

1) Pharmacist wannabee Miss Niagara Falls.
2) World Peacenik Miss Edmonton.
3) Stumbling frenchy Miss Quebec City.
4) College-bound Miss Prince George.

## The Final Questions

**Pierre**: Tell us what you know of the alternate sources of world energy?

**Miss Niagara Falls**: "I know there are many sources of alternate energy such as solar energy — which is being tested in the Arizona desert. It's proved to be very economical in the long run. At first it's very expensive. There are also many other kinds such as uranium and natural gas."

**Pierre**: From your point of view, what makes the science of computers so controversial?

**Miss Edmonton**: "I think the only thing that makes the area of computer science so controversial is that many people think computers cannot work for you. And that you'll be losing your jobs, your opportunities, and that you'll be at a disadvantage with a computer. That's not so. Computers can do very much more for society. They can help disabled people. We can make them work for us, and we can, I think, have a whole better society to live in."

**Pierre**: Comment on the style of work of your favorite artist.

*(I don't speak French very well. Still, Miss Quebec City's description of her favorite art work lasts barely ten seconds. It's obvious whatever she said, there wasn't a whole lot of it. And there's no English version this time. She's still flustered, and it shows.)*

**Pierre**: What events in recent times have changed the course of history?

**Miss Prince George**: "In recent times, being the twentieth century, I feel the loss of the Russian monarch has changed the course of history. If it wasn't for the weaknesses of uh Alexander and Alexandra, we could still have a monarch in Russia. Because of the weaknesses in leading the country, and leading their people, and not understanding their people, they've fallen into a

> lapse of Lenin and Communism. If they had been more strong — if his father had trained him better, Nicholas would not have been thrown over by the communists and communism wouldn't be in the world today."

### And The First Runner-up Is ...

Bilingual commie-hater Miss Prince George.

### And The 1982 Miss Teen Canada Pageant Winner Is ...

World peacenik, computer geek Miss Edmonton. Let's have a round of applause, and that crown from Sarah Coventry .....

### The Moral

If you're going to enter a beauty pageant, girls, go with the tried and true "I'm for world peace" speech. It's an oldie, but a goodie. And Miss Teen Canada 1982 knew that. I much preferred Miss Prince George's nonsensical Russian rant. But who's going to vote against a contestant spouting such uplifting piffle as: "If we work hard and put in a conscientious effort, we can strive to attain, and to learn to work together as a world community and not have to hurt each other. Thank you."

And here I've been making fun of beauty pageants. I apologize.

# 20

***World War II never seemed so funny as it is on tonight's Hogan's Heroes, as Carter impersonates Adolf Hitler ....***

A local Boston station had an in-house series called *Ask The Manager*. The show featured the station manager (a likable middle-aged woman) and some smiling executive hanger-on sitting in comfy chairs, responding to letters from loyal watchers. I loved this show! Not because it was flashy or particularly entertaining, but because it really gave me a sense for viewers' tastes — or at least the taste of viewers who write to TV stations. Funny how the simplest shows are often the best.

Every episode would invariably feature one or two letters about the old sitcom *Hogan's Heroes* (1965-71), which the Boston station had been airing faithfully at midnights with great success. A few letter writers condemned *Hogan's Heroes* for portraying a German prisoner-of-war camp as a place for buffoonery. "World War II isn't something to laugh about" was the oft-repeated phrase of naysayers. To tell the truth, I'm a little uncomfortable when anyone declares a topic off-limits for comedy. Humor takes many forms. If you can't make a subject funny, you're just not trying. I will admit to feeling a little weird laughing at *Hogan's Heroes* occasionally. But maybe that's the fault of the show, not the subject matter.

Most letters were positive, however. The writers didn't see anything wrong with the high-jinks of Hogan, Carter, Colonel Klink, Sergeant Schultz, and the rest of the gang. The supporting cast of John Banner (Shultz) and Werner Klemperer (Klink) was first rate, if a little over-the-top.

We had a German exchange student at our house when I was in my teens, so a special effort was made to keep her away from daily reruns of *Hogan's Heroes*. Though at the time I knew little of the Second World War, I realized there was a certain sensitivity around the subject, no matter what side your country was on. Still, *Hogan's Heroes* is more open to accusations of stupidity than bad taste. The exception to the rule — the Hitler episode. Yes, Hitler appears on *Hogan's Heroes*, or appears to appear. And while it isn't the funniest episode, it certainly is memorable.

Called "Would The Real Adolf Hitler Please Stand Up," the 1967 installment finds Colonel Hogan devising a plot to get photos of German defences out of Stalag 13 without Klink noticing. A standard *Hogan's Heroes* plot turn.

But how do we distract Klink long enough to pull off our latest scheme? Hmmm ... I've got it. Get someone to play Hitler to distract Klink! But who? English womanizer Kinch? Lovable Frenchman LeBeau? Or how about dim-witted explosives expert Carter. He does a dandy Fuehrer with his pocket comb.

So Carter gets the job. And thus the scheme is hatched. Colonel Klink, of course, doesn't know Hitler personally. At a distance, he can't tell a real Hitler from a fake.

General Burkhalter is a different story. Though fat, vain and petty, the German general is not stupid. And he does know Hitler personally. So when Burkhalter arrives unexpectedly at Stalag 13, the prisoners' plan appears in jeopardy. But Carter improvises, using his best Hitler voice to expound loudly on the Fuehrer's hatred of fat generals. Wanting to avoid an incident, the portly Burkhalter leaves the scene without greeting the faux Hitler. Hogan and company foil the Germans once again. Roll the credits.

I think it's the image of Hitler as the mass murderer of this century that makes this episode more unsettling than others. There's no mistaking Hitler's look. He's one of a kind. The greasy, brushed forward hair and little mustache conjure up images of death and destruction — no matter how wacky he is played. Klink, Hogan and Schultz are fictional. Hitler is real.

But that's not to say Hitler can't be funny. During the war years, Charlie Chaplin (1940's *The Great Dictator*), and Ernst Lubitsch (1942's *To Be Or Not To Be*) had great fun with the Fuehrer. Of course, the true devastation of the war had yet to be revealed when those movies were made. Mel Brooks had less luck with his 1983 remake of *To Be*, drawing comparisons to, you guessed it, *Hogan's Heroes*.

Of course, there have also been plenty of well-meaning "dramas" about World War II that were equally insulting to the memory of those who fought and died in the war. Check out 1962's *Hitler* with Richard Basehart, a movie which implies that a sexual dysfunction caused the war.

Hitler can be funny, if he's handled properly. It's just that TV sitcoms seldom have the depth or subtlety to do justice (or injustice) to sensitive topics. But let's not limit ourselves. The definitive Hitler sitcom may be made some day.

The irony of this whole rambling debate is that German television currently airs a dubbed version of *Hogan's Heroes* to an appreciative audience. Called *Ein Kafig voller Helden* (translation: *A Cage Full of Heroes*), the show features dubbed voices (the same guys who dub the German *M*A*S*H*) and a few script changes to make it more "relevant" for today's Germans. And yes, they air the Hitler episode (titled "Film und Fuhrer"). If the Germans don't mind, should we?

***

Over the years, Werner Klemperer has taken a lot of heat for playing the bumbling Colonel Klink. But the actor, who won two Emmys for his *Hogan's Heroes* role, remains unrepentant about Klink. He even reprised the monocled Stalag commander in an episode of *The Simpsons*.

Klemperer was once a guest on *Politically Incorrect*, a late-night comedy debate show. One topic concerned the town of Glendale, California. Apparently, the town council had placed swastikas on a thousand light poles. Jewish groups were outraged, and so was

Klemperer who is a Jew. However, several in the town council contended that the swastika, while a symbol of Nazi Germany, was also a universal symbol of good luck.

I understood this argument, though I didn't exactly agree with it. I first learned of the swastika's place in pre-Nazi history on the fantasy/adventure series *Kolchak: The Night Stalker* (1974-75). If I remember the episode correctly, intrepid reporter Carl was investigating the grisly murders of elderly Jews. It was assumed that the killings were hate crimes, because swastikas were being found near the area of the murders. But Carl soon discovered that the swastika is also a symbol of good luck for Hindus living in the area. In this case, the swastikas were being used by locals as a way to ward off an ancient Hindu demon that they thought was killing Jews.

Klemperer knew this background as well, but he still wanted the swastikas removed from the Glendale light poles. Because he was Jewish, his family had to flee Germany when Hitler came to power. So it was difficult for him to look at a swastika without getting upset. Host Bill Maher pointed out that Klemperer had looked at a lot of swastikas during his *Hogan's Heroes* days. Klemperer's eyes narrowed at this. You could see the decades of defending Klink welling up in him.

"What does that have to do with anything?" he shot back. "It was a satire. I'm an actor."

Fair enough. Maybe Klemperer just got a kick out of getting paid to make fun of Germans. Therapeutic, perhaps? It's not like he was belittling the war, or those who died in it. Just the losers. And that's fair game, isn't it?

# 21

***He's back, but was he ever away? Tune in to The Nation's Music Station tonight for a special edition of* Intimate and Interactive *and an interview with the baddest of all bad boys …***

[*Fade in … Chaos outside the Much Music building on Queen Street, Toronto. Amid the hurly burly of thousands of packed fans stands Video Jock Master T …*]

**Master T**: Welcome everybody to a special night on Much. They've blocked off Queen Street to accommodate the fans who've shown up for our latest edition of *Intimate and Interactive*. [*A cheer erupts.*] In case you've been on the moon, today's show features the first public appearance in several years of the man most closely associated with World War II.

[*The crowd erupts again with applause, jostling Master T.*]

Yes, in the Much environment tonight, we have former German dictator, der Fuehrer himself! Let's go inside for an up-close and personal with Adolf Hitler.

*[The crowd screams its approval as the cameras cut inside to cozy Much space. Teeny-boppers on the street press against the windows, screaming and cooing at every movement inside. A hand-picked audience of young people sits inside, cross legged, in a semicircle around a comfy couch. Big time Much producer Denise Donlon shares the couch with Hitler, looking laid back and rested in a black tailored suit, He's a little gray around the temples, but otherwise healthy and tanned. When the screaming abates for a second, Donlon uses the opportunity to begin ...]*

**Donlon**: Welcome, Mr. Hitler. It's a pleasure to have you on Much. I've watched your work on the History Channel for years.

**Hitler**: It's a pleasure to be here. And thanks. I try to catch your channel on my satellite dish, as well.

**Donlon** [*surprised*]: Oh really? Where exactly do you live?

**Hitler**: Oh now, Denise. [*Laughs.*] That won't do, will it? I can't say on national TV where I live. I've got a few fans and non-fans who'd like to know my address.

**Donlon**: Understandable. You still get a lot of groupies hanging out near your place?

**Hitler**: Yes. Actually, I've seen a lot more in recent months, chanting and marching up and down outside my digs. I love my fans, but ....

**Donlon**: But there's a limit.

**Hitler**: That's right. I like my privacy.

**Donlon**: You say there have been more fans lately? Your career has been enjoying something of a comeback, has it not? Why do you think that is?

**Hitler**: Why does anything happen? I'm trendy again, I guess. There's a lot of young kids who are just discovering me. Everything old is new again. [*Laughs.*]

**Donlon**: As Master T said in the introduction, you are best known for World War II. But that wasn't your first war, was it?

**Hitler**: Oh no. I also contributed to World War I. I didn't get a credit, or anything like that. It was more a learning experience.

You do a lot things to pay the bills when you're young. I spent a lot of nights in the trenches, and working smoky bars before World War II. I was no overnight sensation.

**Donlon**: What was your first hint that World War II was going to be one of those classic wars that people would remember and talk about for decades?

**Hitler**: I think it was Poland, really. That seemed to catch a lot of people's attention — especially in England. They went nuts there. When we toured through Poland, it was our first taste of the big time. My face was on every newspaper and magazine cover around the world.

**Donlon**: We have a very early video of you just before the big break. Can you tell me anything about making it?

**Hitler**: Oh yes. That was shot during my performance at Nuremberg. It probably wasn't my best show. I had a cold, and had been taking a lot of medicine. For a while, it didn't look like I was going to make it. But we had the cameras set up, so the show must go on, as they say. Our director was smart. She really knew how to capture the real Hitler on film. So it came off well, and got quite a bit of attention. I'm proud of it. People forget — this was a new thing back then. Now every dictator has a video. It's all part of the business.

**Donlon**: So let's watch that clip ...

[*Scenes from* Triumph Of The Will. *And then it's back to the Much environment, as the crowd erupts again.*]

**Donlon**: We're back with Hitler in the Much environment. You seem a little embarrassed by that footage

**Hitler**: Oh yes. I just look so young. [Laughs.] And that hair. Wow. [*Crowd cheers.*]

**Donlon**: We're going outside now to Master T, who's got fans with some questions.

[*Outside Much environment. Master T wades into the packed crowd and grabs a screaming teen fan.*]

**Master T:** Hi, what's your name?

**Fan**: Kevin from Kitchener.

**Master T**: And what do you want to ask Hitler?

**Kevin:** Umm, I just wanted to say World War II still rules as the bitchiest war and, umm, it wouldn't have been the same without Mr. Hitler.

**Hitler** [*listening on monitor*]: Why thank you.

**Master T**: So what's the question, Kevin?

**Kevin**: Umm, Mr. Hitler. Uh, what's it like to be famous?

**Master T**: Did you hear that inside? Kevin wants to know what it's like to be famous?

**Hitler**: Well, I don't really know. Other people see me as famous. When I'm not touring, I live like an average Joe. Really. I try to keep everything in perspective. I garden. Watch a little TV. Keep up on my reading. You know — rest and relax. I'm not a partier. You won't read about me in the tabloids.

**Donlon** [*breaks in*]: Well, that's not true. You did sue *The Star* a couple of years back.

**Hitler**: Well, yeah. I had to. They said that my brain was in a big jar on Ronald Reagan's mantel. I had to set the record straight. My brain's still here [*points to skull*], thank you very much. [*The crowd roars.*] Besides, I donated all my money from the settlement to my favorite charity.

**Donlon**: Which is?

**Hitler**: PETA, of course. Fur is murder. [*The crowd cheers its approval.*]

**Donlon**: During the early part of World War II, you could do no wrong. Your name was on everyone's lips, and people couldn't turn on the radio without hearing you. Were there any bad moments that you recall?

**Hitler**: Bad moments? Oh, probably the Russian tour. It started off well. But after a few big gigs everything sort of bogged down. It was badly organized. Several shows were canceled, and we ended up stranded in Stalingrad when our charter plane didn't arrive. The food was pretty bad too.

**Donlon:** You never had much success in the U.S.

**Hitler**: Well, not like in Europe. But I've always had a strong fan base there — mostly in the mid-west.

**Donlon**: We have a caller on the line, Brent from Saskatoon. Yes, you're on Much. Do you have a question for Hitler?

**Brent**: Yes, um. Mr. Hitler? What's it like being famous?

**Hitler**: Umm.

**Donlon**: I'm sorry, Brent. He already answered that. Do you have another question?

**Brent**: Uh, no. No, wait. Yes ... um ... Mr. Hitler. You've been famous for a long time, right?

**Hitler** [*hesitates*]: Yes.

**Brent**: What's that like?

**Hitler**: Just great.

**Donlon**: Thanks, caller. Mr. Hitler, things went downhill for you in the later part of the war. Your popularity plummeted. You also lost a long-time member of the group. What happened?

**Hitler**: You mean Rudy Hess? Well, it was a mutual break-up. Me and Rudy were going in different creative directions, so it was decided that we would part company. I thought we were still friends, but then I started reading in the British press that Rudy was saying all these bad things about me. How I'd fired him, and how the group had sold out. I didn't get it. But I chose to keep quiet about it. Take the high ground, and carry on.

**Donlon**: Things were never the same after that break-up, were they?

**Hitler**: Career-wise, no. There was a noticeable downturn. The British invasion was in full swing, and a lot of Americans were also making the scene. I had decided not to tour anymore. No one was listening anyway. It was too much of a freak show. So I decided to go into seclusion, focus on personal projects, and spend time with my friends and family.

**Donlon**: Did you think of giving up the business?

**Hitler**: No. Never. World domination is my life. I don't know how to do anything else. Well, maybe produce.

**Donlon:** Any regrets?

**Hitler**: I do wish I'd taken more time to enjoy my time on top. But you get so caught up in the business part of your career, that you miss the good stuff. No regrets, though. [*Grins as crowd screams its approval.*]

**Donlon**: Master T? I understand you have someone outside who has a question?

[*Cut to outside Much building*]

**Master T**: Yes. I have Jennifer. She's from Montreal. And she won a trip to Toronto from Much for coming up with the best question for Mr. Hitler. So Jennifer, let's hear it.

**Jennifer**: Fame ... we hear that word used a lot today in the media. What does the word fame mean to you, Mr. Hitler?

[*Cut to inside Much environment.*]

**Hitler**: Ummm. A bad song by Irene Cara?

[*Crowd erupts in knowing laughter*]

**Donlon** [*still laughing*]: Good answer. Umm, anyone other than Irene Cara that you like today? You seem to have influenced a lot of people.

**Hitler**: Oh, I know. A lot of my material is getting covered by young people today. I seem to be more popular dead than alive.

**Donlon**: So you are dead?

**Hitler**: In body anyway.

**Donlon**: Thanks for coming on Much, Mr. Hitler. We hope to have you back sometime soon.

**Hitler**: That would be a pleasure.

**Donlon**: [*looking directly into camera*] Coming up next week on *Intimate and Interactive*, we'll talk to The Spice Girls.

**Hitler**: Oh I love them. They're great.

**Donlon**: Really. Do you have a favorite?

**Hitler**: Sporty Spice.

[*The Much crowd, inside and out, screams its agreement. Fade out.*]

# 22

***Tonight, Kevin must choose between Winnie and the new girl at school on Wonder Years, while a new student exerts a destructive influence on Room 222. And on Death By Television, obsessed TV producer Gerald closes in as fugitive TV reporter Johnston relives his days spent learning the three Rs.***

Schools didn't get much crueler than Aylmer's East Elgin Secondary in the late seventies. Oh, there weren't daily knifings or anything like that. I'm talking mental cruelty. For what can be crueler than having five grades (nine through thirteen) sharing the same hallways? You can see the problem right away. Thirteen-year-olds cohabiting with nineteen-year-olds? Puberty meets adulthood? It was a scary place.

East Elgin High was a bigger-than-average school, boasting two gyms, three floors of classrooms, a decent football team, and some enthusiastic teachers. There was also a big cafeteria and, during my first year, a jukebox. But that ended when someone kept playing Boston's *Don't Look Back* and The Beatles' *Helter Skelter*, disturbing nearby health classes. Goodbye jukebox. For years to follow, student politicians got elected on a pledge to bring back the jukebox. It never happened.

I arrived at the school in 1979 as a typical scrawny grade niner. My hygiene was suspect, I had glasses and a retainer, and my hair stuck out at my ears, giving me the appearance of a koala bear. I wore nothing but jeans and t-shirts with sports logos or names of rock bands on them (The Boomtown Rats, April Wine). I had no skill whatsoever when it came to basketball, football, or baseball. I was one of those guys who passed health and phys ed by looking determined while on the court or field. Teachers love to see you "try" even if you suck, and I knew it. So I faked enthusiasm, and then beefed up my marks in the health class portion.

"Ian put in a good effort this term," would be my gym teacher's remarks. I'd sometimes pull off a B. Maybe even a B-plus. All without talent. Good marks are easy for a gym conman. Still, I wasn't a complete sports washout. I was better-than-average at cross country running, participating in a number of county-wide events. But that's not exactly a big spectator sport, is it? Yes, I was out of the sports and social loop. No doubt about it. And lacking the supportive peer group that all grade nines needed.

The grade thirteens were moving into university mode and didn't pay attention to anyone outside their grade. Most of the cruelty came from the grade elevens and twelves. They were beginning to feel their power as seniors, but were still young enough to get a kick out of picking on defenseless freshmen, if they gave them any attention at all. For grade nine girls, there was the opportunity of dating someone in grade eleven, or getting invited to older kid parties. But nobody wanted to have the slightest thing to do with grade nine guys. We were useless creatures who were expected to stay out of sight for a couple of years.

Small town uselessness is worse than big city uselessness. What's there to do on a Friday night? There's not even enough activity at the local convenience store to make hanging out there worthwhile. I went to a few school dances my first year, spending much of my time drinking pop and leaning against a wall. But the majority of my Friday nights seemed to involve wandering around the streets of Aylmer. Or watching television, which probably explains my career choice.

The closest equivalent TV character to me in grade nine was Marshall Blechman of the short-lived sitcom *Square Pegs* (1982-83). The show starred Sarah Jessica Parker and Amy Linker as Patty and Lauren, hopelessly out-of-touch high schoolers always trying to be a part of the "in-crowd." Why anyone would want to be a part of that group, I couldn't figure out. Most of the folks in the *Square Pegs* in-crowd (including Jamie Gertz as preppie cheerleader Muffie ) were slow, and downright mean. Still, Patty and Lauren constantly pursued in-crowd acceptance. When that inevitably failed, they found themselves hanging around with the dim-witted rocker Johnny Slash (Merritt Buttrick), and fast-talking jokester Blechman (John Femia).

Blechman was a product of the TV age, reaming off obscure references to old TV shows, and doing passable impressions of the celebrities of the day. The fact that no one listened or appreciated his efforts didn't faze him. Blechman was colossally out of step, but oblivious to what others thought. Still, Blechman was happy. He shouldn't have been. But he was anyway. I had the character's sense of humor, but I never reached the level of Blechman obliviousness I wish I had. And though I mocked Patty and Lauren's pursuit of in-crowd acceptance, I too was caught up in my in-crowd dreams.

Still, it was only a dream, for I did little to attain the goal. Indeed, I constantly hurt my social standing by doing the supposedly "nerdy" things — like studying, getting good marks, participating in plays, and dressing up as Radar on *M*A*S*H* day. Why did I wear a cute green hat and carry a teddy bear when I could have just grabbed a housecoat and a martini glass and been Hawkeye? Too obvious, I guess. Stupid. Stupid. Stupid.

Grade level wasn't the only determiner of social status at my school. Aylmer was a small town by any standards, but there was a definite cultural rift between the town students and the out-of-towners, who were mostly farm kids. The third group were the members of the growing German/Mexican community, who tended to be keep-to-themselves country folk with large families. Funny thing. While TV and movies were giving me one Mexican stereotype (big mustache, poncho, outrageous accent), reality was giving me something altogether different. I grew up thinking that being "Mexican" meant people had blond hair, blue eyes, and names like Abe, Henry and Cornelius. Which is true too, at least in my town, anyway.

I hung out with a townie group I'll call the Naysayers, which sums up our approach to life. On the social scene, we were hated by some, liked by some, and ignored by most. Looking back, I can't figure out why the Naysayers stuck together. We didn't seem to get along. Apart, we were nice and relatively well-adjusted. Together, we were uptight and cruel, turning on each other at the slightest provocation. It was tough to even have a conversation. Success on any level was ridiculed by the entire group. So was failure. So was inactivity for that matter. And if you got a date, well watch out. You're gonna get heat from everyone else. The Naysayer Prime Directive — don't do anything different, lest you want to take heat.

One of our obsessions was homosexuality, or the appearance of homosexuality. We were as tolerant as your average grade niner (which ain't saying much). Still, we liked to accuse each other of being gay — which apparently was the ultimate insult of the day. The worst thing you could do was to use a word or phrase that implied anything to do with homosexuality, even in the vaguest, most cryptic way. Of course, using the word "gay" or "queer" in any context was a no-no, as were seemingly harmless phrases like "I beat him good" or "You're going down, man." It was all very *Beavis and Butthead.* Homosexual double entendres lurked everywhere in our world. It was safest just to keep quiet.

It's a shame we wasted all that energy attacking ourselves. For grade nine is all about survival. Getting through the year alive is the only realistic goal. We were like sailors on the *USS Indianapolis*, which was sunk in the Pacific in the dying days of World War II. The only way they survived the shark-infested waters was to form small groups, and beat the water when a fin loomed. Had The Naysayers worked together like the crew of the *Indianapolis*, things might have been OK. But instead we turned on each other and were eaten. Metaphorically speaking.

# 23

***On tonight's Adventures of Ozzie and Harriet, Ricky comes home from school to confess that ... nothing in particular happened that day. Enjoy ...***

Don't trust TV. I don't.

It's a part of me. But there's no law saying I have to trust that part. Accept it. Yes. Trust it? Uh uh. No way. TV is an entertainment vehicle, powered by rigid time slots and the need to change the topic before viewers grow bored. Television is life with the dull and subtle parts cut out.

TV teaches you how to deal with a terrible tragedy, but gives you no idea how to cope with the mundane bits of daily life. I know how to foil knife-wielding criminals in dark alleyways, and the proper method of sweating a confession out of a reluctant murder suspect: "Hey! While you're sitting here all quiet, your buddy's in the other room singing like a canary. He's giving you up, man. Right now, he's telling another detective how it was you who had the gun. You! So are you gonna be the smart guy and tell me what happened?

Or are you gonna be a chump, and let your buddy walk while you rot in a hole for twenty-to-life? What's it gonna be, smart guy?"

Yet nowhere in my vast hours of TV watching is there any info on how to deal with less dramatic situations that arise out of the average get-up, go-to-work, come-home, have-dinner, go-to-bed life. Even in something as gentle and low-key as *The Adventures Of Ozzie and Harriet*, the mundane is strangely absent. Ricky always has some problem with girls, school, or zits. And Dad and Mom solve their son's problem within the half-hour time frame (with commercials), sending their pompadour-wearing progeny off to bed to dream of becoming a rock star.

But we never see Ricky come home with no problems, sit down in front of the TV, have dinner, talk on the phone, do his homework, then go to bed. Where's Ozzie and Harriet's advice on living the uneventful life? On the other hand, where are all those teenage problems that take more than a half hour to solve? That's the flaw with TV as a teaching tool. It's dependent on good theater — whether it's a sitcom, a one-hour medical drama or a newscast. Viewers want their excitement in bite-sized pieces.

Not that I want television to reflect all aspects of our lives. Heck, if real life were made into a TV show, it'd be canceled after one episode. We live in an unedited world, a twenty-four-hour a day routine of non-drama, inaction, and body functions. No climaxes, no foreshadowing, no ominous background music to cue us when something bad is going to happen. There are no flashbacks or opening summaries to remind us of the key players in the on-going drama, or what exactly the on-going drama is. And real people certainly don't go through dramatic "arcs" on their way to self-discovery.

Our lives are bad TV — a low budget, seventy-two year miniseries with little plot, few highlights and tragedies, and way too much character development. Ninety-nine per cent of the characters who do make an appearance end up having no bearing on the central story. Try pitching that concept to a network, and see how far you get. I don't expect to see it on the fall schedule any time soon. And if it does show up, about six billion people have a good case for copyright infringement.

Of course, maybe all we need is a good theme song. Hmmm ....

***

Now that I think about it, maybe there have been moments in my life that resembled TV. I can think of one bit of past reality that lends itself to a TV movie dramatization. With a tweak or two, it could be a less-preachy After School Special. Or maybe one of those low-brow Disney TV movies that air early Sunday night, minus the talking cars and men in drag.

**The Title**
*The Campaign*, AKA *I Was A Teenage Propaganda Minister*

**The Theme Song**
*AC/DC* song "Let's Get It Up."

**The Credit Sequence**
JACKIE EARLE HALEY as Cliff.
KIRK CAMERON as Golden Boy.
MERRIT BUTTRICK as The Singer.
JOHN VERNON as Principal Wormer
WILL GEER as The Grandfather.
And introducing JOHN FEMIA as MARSHALL BLECHMAN as IAN JOHNSTON
PRODUCED And DIRECTED by IAN JOHNSTON
Adapted from the book *Death by Television*, by I. Paul Johnston, esq., BA, BJ.

## The Prologue (AKA The Annoying Part Where The Premise Is Hinted At, But Not Revealed)

In high school I usually hung around with a group I now call The Naysayers. There were a few exceptions, though. Cliff was a farm boy whose main obstacle to mass acceptance was a bad case of high school acne. Still, he did have an impressive teenage beard, and a sense of humor similar to mine. I got to know him in grade nine, when we worked together to pass our accounting class. He watched the door while I swiped the answers from the teacher's filing cabinet. What a team.

But I'll always remember Cliff for that time in grade twelve when he had a vision — to unite for one brief moment those who saw themselves as geeks, outcasts and put-upon grade niners. Well, maybe that's my hindsight version of Cliff's vision. I think Cliff just wanted to be a shit-disturber. Either way, it was a fun week...

### The Heroes

Over the high school years, Cliff and I slowly moved up the pecking order. By grade twelve, we weren't exactly considered cool. We were more like Wally Cleaver in the last couple years of *Leave It To Beaver*. Though considered too "square" to be really popular, we had earned a grudging identity as "OK guys" from the snooty, fashion-conscious, in-crowd set, the lowly freshmen, and some of the folks in-between. We were normal, middling guys, who knew from experience the pain of being perceived as nerds, geeks, or minor niners.

### The Villain (AKA The Necessary Bad Guy For A TV World Obsessed With Stereotypical Characters)

Somewhat untouched by all that usual high school hierarchy stuff was the son of a successful tobacco farmer. He wasn't part of an in-crowd exactly; he was one all by himself. He had classic good looks, and knew it. If he was full of himself ... well ... he could get away with it. He got good marks, was quarterback of the senior football team, captain of the senior basketball team, and fronted the only rock band in our high school. As a sophomore, he'd even been elected Vice President of the Students Council, almost unheard of for a guy that young.

Golden Boy wasn't even the typical football jock. He hung out with other jocks, but didn't seem to get off on the same shenanigans — like punching younger, weaker kids in the arm, or offering up wedgies to unsuspecting freshmen who'd carelessly let their underwear hang out. No, Golden Boy was above all that goofy stuff. He was the young-man-who-would-be-king of East Elgin High. With his year of service as Vice President, it was expected Golden Boy would eventually complete his rise to power, becoming student prez for his final year.

Cliff had a different idea.

## The Premise (AKA The Part Where It's Finally Revealed What This TV Tale Is About)

Cliff couldn't stand the idea of Golden Boy getting a free ride into the presidency. So he entered the election at the last minute and asked me to help. The plan was to approach the election as something of a joke — a Keep-The-Other-Guy-On-His-Toes campaign. Just to make G-Boy sweat a little on the way to the presidency.

Our strategy was simple — Be weird. At best, the voters would remember Cliff's name on election day. Candidate Cliff played it straight, talking to people on an individual basis, showing all those skeptics that he wasn't totally nuts. He was an okay guy. By high school standards, of course.

I was Cliff's clown prince of propaganda, taking all the election flack, so Cliff could stay in the background, working the phones. I organized impromptu rallies, carried banners through the school, and conducted goofy singalongs of rewritten jingles from old American presidential campaigns.

"I like Cliff / You like Cliff / Everybody likes Cliff."

## The Election Trail (AKA The Part Featuring Montage of Election-related Scenes Combined With The Latest Pop Hit For A Music Video-type Effect)

Poland had been in the TV news for months, what with its political and labor strife, so our first poster slogan reflected that: "Solidarity Forever." It didn't mean anything, but it sounded good, in a Marshall Blechman/*Square Pegs*/absurdist way. And I figured there were benefits in using a phrase that appears on TV every night, even if it has no real connection to high school politics. Still, "Solidarity Forever" does make you think. And when you're thinking about a candidate's slogan, you're halfway to voting for him.

Eventually, we created a cover story for that slogan. Everyone thought Cliff was Polish. He wasn't, but why fly in the face of public opinion? So "Solidarity Forever" became a reflection of Cliff's Polish immigrant roots. Make fun of our poster, and you make fun of Cliff's family, man.

Our second slogan was equally Blechman-like: "Cliff — The Only Alternative." Without actually saying it, we were telling the voters that if they didn't want to vote for the Golden Boy, their options were severely limited. Vote for the quarterback or vote for Cliff. Vote for the guy who appears to have everything or the normal, average son of Polish immigrants. Those are your choices.

Cliff clicked immediately with the school populace, though I'm not sure it was all our campaign's doing. For it was at this point that I learned the awful truth about the true nature of high school ...

## The Revelation (AKA The Middle Part Where The Heroes Learn A Valuable Lesson)

It all came clear to me as I was working the crowds during election week. I'd approach some voters, and out of their mouths would pour all this petty anger aimed squarely at Golden Boy. Though they didn't know him at all, students seemed to hate the guy just because he was Golden Boy. Even members of the in-crowd expressed irrational hatred of the guy. He hadn't done anything to them; they just didn't like what he represented.

Maybe I was just stupid. But I had always figured members of the in-crowd were confident, well-adjusted people. Yet, as I talked to these potential voters, I was left with one impression: we're all geeks. At least, we think we are. For inside every East Elgin High student were troublesome feelings of inadequacy and paranoia. No matter how apparently cool or successful (socially or academically), we each saw ourselves as nerdy in some way. The in-crowd weren't really the hippest or the best-looking, as I had thought. They were simply the ones who aspired to that status with the greatest fervor. Of course, they could never attain that ideal. So they were doomed to failure, paranoia, and self-confidence problems, just like the rest of us.

And what better way to address your feelings of inadequacy than to lash out against the one guy with everything going for him? Why shouldn't a self-loathing high school geek or nerd hate a smart, well-spoken, well-off, nice-looking, football and basketball-playing rock singer who dabbled in student politics? And who wants to vote for a guy who has everything including impeccable qualifications?

So without doing anything wrong exactly, G-Boy became the focus for irrational teenage hatred during that campaign. I even got the impression his own football team had turned on their quarterback in the election's dying days. The campaign had taken a decidedly unexpected turn ...

### The Conflict (AKA The Fabricated Part Where The Heroes Articulate Their Feelings For Dramatic Purposes)

**Ian**: It used to be about the students, man. Now it's about you.

**Cliff**: So now you're turning against me? Well, thank you, Judas. Why don't you just go help Golden Boy?

### The Resolution (AKA The Made-Up, Clichéd, Tear-Jerking Part Where The Heroes Again Articulate Their Inner Feelings For Dramatic Purposes)

**Ian**: I love ya, man.

**Cliff**: I love you too, man. Now, let's win this election together!

**Ian**: You said it! Solidarity Forever!

### The Climax (AKA Speech Day, AKA The Part Where The Exciting Stuff Happens)

Speech Day concluded the campaign. It was a double header — beginning with an assembly for the junior grades (nine and ten), and concluding with the seniors (eleven through thirteen). I preceded my man with a short intro designed to work the crowd into a geek frenzy in anticipation of Cliff's arrival.

"Cliff! A man of few words — most of them Polish," I began, eliciting a huge cheer from the assembled mob, a good many of whom were well-placed ringers with signs and placards I'd written myself. Yes, it was a real (and fabricated) Cliff love-in, and I was at Ground Zero. To add to this assembly mania, I'd gone behind Golden Boy's back and hired his own band to play for Cliff at the assembly. Without G-Boy of course.

The band agreed without a fuss. I guess they'd also fallen victim to the "Golden Boy Has Everything" negative vibe sweeping the school. Filling in for the gilded one on vocals was a guy so far down on the high school social ladder as to be undetectable by radar. The Singer was an easy-going sort who had steel wool hair, a middle aged paunch, and a penchant for sweating profusely. All the time. No matter the temperature. But though a social outcast, The Singer had earned notoriety in recent weeks for his throaty rendition of the AC/DC song "Let's Get It Up." AC/DC was the current fave on Crop-Tours — our quaint, high school, tobacco-farm-related euphemism for driving around in the countryside and drinking beer. You couldn't stand at Aylmer's main intersection for five minutes on a Friday night without the sound of AC/DC's "Back In Black" blaring out of a passing van. So our music act seemed foolproof. Heck, the title "Let's Get It Up" was even an obvious sexual double entendre that would secure backing from teenage homophobes and Beavis and Butthead precursors.

So The Singer growled and sweated his way through the tune as the crowd went wild. And as the song reached a volume peak, a loud chant of "Cliff! Cliff! Cliff!" arose in the cafeteria. I'd never heard such volume. Not even from the long-gone jukebox. This was cultish stuff. And at just the right moment, a dozen minions walked on stage sporting three-piece suits and carrying our leader aloft in an easy chair. It was a Polish Godfather sort of thing.

With his entourage behind him, Cliff launched into a speech that was funny, self-deprecating, and lasted all of forty-five seconds — three qualities I admire in speeches. But then again, I wrote that masterpiece.

Golden Boy was none-too-amused. Defeat suddenly seemed real to him. So before the second, senior assembly, he talked to his band, convincing them to back him on a song that he hoped would prove that he was also goofy and fun like Cliff. I can't remember whether he sang "Heartbreak Hotel" or "Hound Dog." It doesn't matter. The important thing was that it was an Elvis song. In 1982, Presley was only five years in the grave, and still remembered as a greasy fat man by Aylmer teens. In an early eighties confrontation between Aussie rockers and dead guy Elvis, AC/DC wins.

## The Celebration (AKA The Part Where They Play Some "Ode To Joy"-type Song While A Montage Of Teenage Revelry Unspools)

More than six hundred students voted in that election, twice the number of the previous year. A ballot insider put the final tally at five-to-one in favor of Cliff. Student self-loathing had spoken, and Cliff was their man. Even Cliff's younger brother got in as secretary on the coattails of our landslide. Not just a victory for Cliff, but for Polish immigrants everywhere.

We had a great victory party at Cliff's farm. Even an impassable access road couldn't stop the masses from showing up. G-Boy's band played in the barn loft, and everyone got drunk and tossed beer bottles through the panes of a nearby greenhouse. Cliff's mom and dad were none-too-pleased with the destruction and mayhem. So early the next morning, Cliff's dad roused the campaign team to stack cords of wood at some mosquito-infested backlot. A dose of political reality and accountability.

## Epilogue (AKA Where Are They Now?)

In one of his first acts of office, President Cliff attempted to bring in a smoking room (complete with new furniture) for nicotine-addicted students. He even got his plan by the council's teacher rep before the principal stepped in to nix the whole idea. Student councils only have as much power as the principal allows. Cliff took the smoking room incident as a signal to lose interest in his job. Maybe he should have just stuck with tradition and tried to get the jukebox back in the cafeteria.

Still, Cliff didn't swear off politics entirely. Last I heard, he was running for a council seat for Ward 3, Bayham County. Freed of political baggage, Golden Boy went on to quarterback the football team to some sort of regional championship that season. Last I heard, he got married, had kids, and continued in the family business. As for me, why I went on to form a media conglomerate, buy the Atlanta Braves, and sail the U.S. to victory in the America's Cup.

## The Closing Theme

A Polish folk tune (with accordion) to reflect Cliff's immigrant heritage. Segues into a reprise of "Let's Get It Up" ...

## The Closing Disclaimer

The story you have seen is based on true events. Some characters are composites of real people. Some real people were composites of fictional characters. Several incidents (the best ones) were fictionalized and dumbed down for dramatic TV purposes.

No animals were harmed during the making of this production. Come to think of it, several horses did give their lives to provide glue for the campaign posters used in this teleplay. But they're just horses, so who cares, right? Dumb, mean animals.

***

High school year books are a lot more amusing fifteen, twenty years after the fact. At the very least, the hair and clothes are worth a chuckle. My 1983 yearbook also chronicles one of the funniest scheduling errors of my senior year. For some inexplicable reason, Photo Day was held on Slave Day, when senior students dress up as slaves and auction themselves off as a fund-raiser. Didn't anyone see the conflict?

So in the group photos of the Drama and UN Club, there I am dressed in rags with a heavy chain around my neck. All those nice clothes mom bought me, and this is the thanks she gets? Oh well. At least I wasn't like my friend Gary, who appears to be dressed as Aunt Jemima. He's hiding in the back row, but I can still see him.

My grad photo comes complete with typical yearbook remarks that must have seemed hilarious to me at the time:

> Favorite Phrase: "*There's a signpost up ahead. The next stop — The Twilight Zone.*"

That's to be expected. Nice to see I didn't imagine that whole tobacco, Twilight Zone thing.

> Career Goal: "*Propaganda Minister for the Fourth Reich, led by Mein Fuehrer Cliff. Seig heil!*"

Whoops. That doesn't play in the 1990s. It was a school boy joke, of course, made just after our political victory. I thought it was pretty funny at the time. But I now suspect that quote pretty well excludes me from any chance of running for political office:

> **Reporter:** Mr. Johnston?
>
> **Johnston:** Yes?
>
> **Reporter:** Can you confirm published reports that claim you made several Nazi references in your high school year book?
>
> **Johnston:** Well, it was a joke. There was this election see, and I helped run the campaign and, well, uh ... no comment.

My friend Cliff was much smarter in his 1983 grad photo, listing his career goal as simply "ruler of the world." Man, why didn't I think of that?

# 24

***On tonight's Full House, Bob (click) ... and don't miss Tim Conway's laugh-a-minute guest appearance on Married With (click) ... Guzman steps in, waits for the sign ... and delivers — ball two ... (click) ... I feel like chicken tonight, chicken tonight ...***

The remote control has changed my universe. As a single guy, I have always maintained full control of my massive, 47-button clicker. And over the years, I've evolved into a channel-clicking mad man who seemingly can't make up his mind what the hell to watch. TV programs used to be interesting. They still are, but not interesting enough to watch all the way through.

During baseball season, I try to watch the Blue Jays while keeping an eye on one or two prime-time shows. To make it even more hectic, I also keep a video tape cued up for those rare moments of simultaneous network commercial breaks. It's a bit chaotic, but baseball can be slow going, prime-time shows have commercials, and I have to use every opportunity to attack a stack of screening tapes that isn't getting any smaller.

But I'm not out-of-step with the rest of Canadian viewers. Check out a 1997 survey from the Toronto-based opinion and market research folk at The Strategic Council. In it, one thousand people in the five urban centers of Halifax, Calgary, Toronto, Vancouver and Montreal were asked how much TV they watch, and just how they watch it.

The results show that 22 percent of Canadians eat dinner in front of the TV every night of the week. That figure was much lower in my home town of Halifax (13 percent). No crumbs on our couches. I'm guessing the discrepancy between country and city is caused by the fact that Haligonians don't spend as much time going to and from work, so we have more time to put together a decent meal at the dinner table. Or perhaps TV tray technology is weaker in the east.

Canadians also have itchy trigger fingers. According to the survey, 26 percent of respondents say they watch two programs at a time, while 64 percent watch one. In my town, the multi-channel watchers came out on top, with 42 percent of respondents watching two programs, and 33 percent watching one. Methinks a third of you aren't trying hard enough.

As for what we watch, news (31 percent) and movies (30 percent) remain the top choices throughout Canada. Sports and educational programs finish out the top four.

So what does this all mean for me on the east coast? According to the survey, the majority of Haligonians are channel-flipping, news-loving, movie-watching psychos who don't like to mix food consumption with tube viewing.

Hey! That's not fair. I like eating in front of the tube.

# 25

***In a special, Sweeps Week installment of Death By Television, a false media report forces fugitive TV reporter Ian Johnston to defend the very medium that caused him to flee …***

Television is an ultra-violent world rife with profanity and nudity, a hyper reality that celebrates criminal behavior.

This certainly seems to be the media's opinion of TV. Even television news appears to go along with this TV-as-bad-influence opinion, essentially condemning itself. Of course, it's easy to attack TV because it's a *thing*. It's not a person or an institution. It's TV: faceless, and thus easy to sneak up on.

Now, I don't want to become the defender of TV, but I don't see TV as violent and evil. For me, the exact opposite is true — television is too good and too clean. It presents us a ridiculously moral world where evil rarely (if ever) goes unpunished, where drinkers and smokers are social misfits, and where problems fade with the final credits.

We make demands of television that we don't seek in our own lives. Television is how we wish life was, but can never be. Think about it — how many lead TV heroes are smokers? Can you name one? The American spin-off of the British series *Cracker* had a lead guy who smoked a bit, but that show failed in its first season. And the *Cracker* dude isn't exactly a role model anyway. Heck, the only time people smoke on TV is to show that they're unglued or idiots. On *Beverly Hills 90210*, Brenda started smoking in one episode, and the whole teeny-bopper gang called her names till she quit. Talk about peer pressure.

*Cheers*' Rebecca didn't pick up the habit (again) till the writers decided to change her character from a no-nonsense business woman to an insecure bonehead. And let's not forget *The Simpsons*. Marge's chain-smoking sisters are the constant butts of humor. Remember when their habit causes an unborn Lisa to cough in the womb? Yet, if statistics reflect reality, 31 percent of Canadians smoke. And not all of them are psychos, idiots, and off-kilter women. At least I hope not.

As for TV violence, just how many violent crime-shows conclude with the bad guy getting away? Most certainly the average is a lot less than in real-life crime. Even the arrival of grittier shows like *NYPD Blue* and *Homicide* (where the bad guys succeed occasionally) hasn't evened the score. No, life in the TV-crime world remains like that in the original *Perry Mason* (1957-66). The TV lawyer only lost one case in nine years, and that verdict was eventually reversed. That's 271 cases won, none lost. A pitcher could go a long way with that record in the majors. So could a crown prosecutor, actually.

Hurting TV's image have been a bevy of studies and surveys into TV violence. Though some have been valuable and comprehensive, a number have been high-profile nonsense, breaking television down into little bitty pieces, and reducing shows to "incidents of violence" or "spoken profanities." But these studies don't allow you to calculate the context of a violent act on TV, nor can you accurately chart why a character used profanity.

We in the media don't help with our reports on TV violence studies. We love writing stories on TV surveys. Everybody knows what TV is all about. And these surveys offer easily translatable statistics that lend themselves to news stories and attractive headlines.

The one that bugged me arrived in 1995. Conducted by several U.S. universities, the *National Television Violence Study* pointed a finger at the rise in TV violence. Or so it appeared. When the study was published, it created a media stir in its "revelations" on the state of TV. One oft-quoted survey statistic asserted that in 73 percent of TV shows, the violence goes unpunished. Sort of blows my TV-as-a-moral-fantasy-world theory, eh?

Not quite. Even when this story was big news ("Violence Goes Unpunished On TV — Study" screamed the headlines. "Crime Does Pay On TV" was the lead on the TV news), I was suspicious. So I sent away for the study summary to get the fuller story. Upon closer examination, the oft-quoted violence stat actually asserted that in 73 percent of *scenes* with violence, the violence is unpunished at the time it is committed. That's one scene. Not the whole show.

Which makes sense, eh? How often do you see a bad guy murder someone, then get arrested before he leaves the scene? Pompous rich dude Patrick Macnee murders his philandering wife, and *Columbo* catches him before Macnee wipes his wife's blood off his hands? Heck, if TV punished the bad guys as soon as they committed a violent act, we'd have sixty, one-minute dramas in a one hour show. Or is that forty-four dramas plus commercials?

Deeper in the study was the real truth: 62 percent of the bad characters on TV are *eventually* punished for their violence. So statistics show that two-thirds of bad guys get their come-uppance on TV, not 73 percent go unpunished. Makes a difference.

Other freaky stats fly in the face of apparent reality in the United States *National Television Violence Study*. The report made a big deal out of the fact that only 15 percent of good guys are punished for their violence. Of course. Why would a drama punish a good guy anyway? Usually, he's only acting in self defense or to stop a bad guy. If a good guy is going out and busting heads without just cause, he no longer qualifies as a good guy, does he? Where's the dividing line between good and bad guy? And who makes the call?

***

It must be noted that this violence study is not necessarily talking about TV shows produced by networks. To tell the truth, most of the complaints about the so-called TV violence are aimed at feature films airing on local channels.

Another study — *The UCLA Television Violence Monitoring Report of 1994-95* — found that only ten of 121 prime time shows used violence frequently or questionably (questionably meaning that it's used in a heroic or graphic manner). In contrast, the same study found that among 118 big screen movies that aired on TV, fifty featured questionable images of violence.

Apparently, theatrical movies airing on TV are more likely to contain offending violence than TV programs produced by the networks. This makes perfect sense. The networks have predetermined rules on what they can and can't do. A PG rating is about as bad as TV programs tend to get. Movies have no such restrictions. Even R-rated flicks can air at 8 P.M. (with a few minor cuts), or later at night with none. Of course, a Bruce Willis or an Arnie Schwarzenegger actioner is still more violent than the most violent TV shows, even with edits and dubbing for inappropriate language. "Mother freaker" is a very popular swear word in Bruce Willis movies airing on TV.

TV stations could reduce TV violence complaints by doing a better job of previewing the flicks, editing out the offending bits, and airing the raunchier, violent movies later at night. Not that I personally advocate censorship or anything. Maybe this whole issue would be easier if viewers just did some advance research on what is offered on TV. Don't just turn it on and start surfing. Become educated.

Aww crap. I've gone and spoiled a beautiful thing like TV by assigning homework.

***

At risk of hurting my argument, I must note that the *UCLA Television Violence Monitoring Report* lists *America's Funniest Home Videos* as one the most violent shows on TV. I guess that makes sense. In 100 percent of scenes featuring a kid hitting his dad in the groin with a baseball, the kid is not punished for his crime, but is instead awarded $10,000 by Bob Saget.

# 26

***The reporter fugitive reflects on a life of inactivity in a navel-gazing installment of Death By Television.***

Who better than me to serve as guinea pig in testing the long-term effects of TV violence? Violent American shows like *Mannix*, *Hawaii Five-O* and *McCloud* have been pouring across the border and into my little TV-watching brain since I was old enough to sit up straight and watch. Maybe before. I have photos of myself as a newborn lying on the couch right in front of the old floor model ...

According to the latest annual Statistics Canada surveys, the average Canadian clocks in about twenty-four hours a week of TV-viewing. Erring on the conservative side, I estimate I watch thirty hours a week. I watched less than that when I was two, but I watch more now. So thirty hours seems a fair figure. Thirty times fifty-two weeks equals 1,560 hours. Multiply that by thirty-four years and you get ... holy cats! 53,040 hours! That's ... that's .... six years!

That's equivalent to all of high school and my first year of university. That's more than my entire existence in the 1960s. That's more than a sixth of my life. Gosh, I'm glad I get paid for watching television. I just depressed the hell out of myself.

Here's an historical perspective on what has been accomplished in the same time (or less) than I've spent watching TV:

1) World War II.

2) World War 1.

3) The release of every Beatles album from *Meet The Beatles* to *Let It Be*.

4) The publication of Shakespeare's *Richard III*, *The Taming Of The Shrew*, *Titus Andronicus*, *The Two Gentlemen Of Verona*, *The Life And Death of King John*, *A Midsummer Night's Dream*, *Richard II*, *Love's Labor's Lost*, *Romeo and Juliet*, *The Merchant of Venice*, *Henry IV*, (parts one and two), *All's Well That Ends Well*, *Henry V*, *Julius Caesar*, and *Much Ado About Nothing*.

# 27

***A trip through a small town places the fugitive TV reporter at the scene of a terrible train accident in this week's Death by Television. Viewer discretion is advised.***

Despite my bluster, I worry sometimes: has all this TV watching made me blood thirsty and desensitized to life's true violence? Hard to say. I did have an interest in horror movies for quite a few years, but that's not much different from guys into heavy metal music. I don't get into fights. As a coward, but a sly one, I've done a great job of avoiding them. The last one occurred when I was twenty-one. The old boyfriend of a girl I went out with cold-cocked me outside a 7-11. He was mad that I went out with her after she broke up with him. Yeah, I didn't understand it either. Heck, she was sitting in the car while he was beating me up, so what's the problem here? You win. Let me buy my two liter Pepsi, and I'll be on my way. Still, it wasn't much of a pummeling. My glasses flew off on the first punch, so I was basically defenseless. I got a shiner that bothered me a bit.

As for my reaction to scenes of violence in real life, the jury's still out. Real violence is rare in our lives. And how we react to it is a very personal thing. And the only real incident of violence that comes to my mind is so extreme as to not be particularly instructive. And besides, I didn't see the violence. I just saw the aftermath. It's an interesting story anyway ...

I was working at a small town weekly newspaper. It was a summer student job, and a pretty good one. Basically, I was responsible for writing news stories and taking and developing photos, though I had little or no experience at either job. Scares the hell out of me just thinking about it now.

It was a slow Monday. And with the Tuesday deadline looming, I was short of stories. In cases like this, I was taught to grab a camera and hit the streets in search of local colour to fill the paper. Maybe a story. Maybe just a photo. Both if I was lucky.

I knew I'd scored big as soon as I reached Main Street. Far down the road by the railroad tracks, I saw flashing lights. Even if it was a minor accident, I'd still have a photo. This being a weekly newspaper, I take what I get. When I got to the scene, I didn't see anything. Cops were putting up barriers to keep the growing crowd back, but from what, I couldn't tell. So I asked a local merchant, who informed me rather matter-of-factly that "Cheetah had finally done it."

Cheetah was the nickname of an ill-kempt street guy who wandered around the town generally not bothering anyone. His big claim to fame, till that Monday, had been a human rights case against a local restaurant which had banned him for various eccentric behaviors, possibly related to his epilepsy. That incident, eventually carried in the national media, made Cheetah a quasi-celebrity. And he took to wearing a hand-scrawled sign around his neck, claiming discrimination by other restaurants.

Cheetah also occasionally told folks he was going to kill himself, often mentioning the railway tracks as a final destination. He put that boast to the test that Monday morning, lying down on the track as a train moved slowly through the center of town. That's why I didn't see a body at first. The large Cheetah pieces had been covered with tarps. The smaller bits, already drawing flies, really didn't look like much of anything. I'm being straight forward, "just-the-facts-ma'am" because there's no way to exaggerate this surreal circus.

I would have been satisfied just to watch from behind the police barrier with the rest of the curious town folk. But a local cop recognized me, and motioned me forward to join the coroner and the clean-up crew. I snapped a few photos (most of which were too gruesome for publication), and tried to stay out of the way. My heart was racing. But I wasn't disturbed, or repulsed. If I sound callous or uncaring, it's not on purpose. It was just so surreal. So unlike anything I'd experienced. The crowd felt the same. There were no public displays of sadness or nausea. Everyone was just, well, baffled.

The arrival of an ambulance added to the strange scene. The attendants emerged and calmly donned plastic gloves. Up and down the tracks they went, picking up pieces like lumps of damp sod, depositing them in an increasingly bloody sack. Then came the fire truck, and we all had to step back while they hosed down the train. Developing the photos later, I got a little creeped out. But I wrote the story, and later enthusiastically recounted the details to anyone who would listen.

I was more disturbed that the event hadn't bothered me as I assumed it should. I'd seen the shows on TV where people are weeping and screaming at scenes of violence. Or they experience nightmares in which they wake in the night, soaked in sweat.

None of that happened to me. Desensitized? Maybe. But still, I remember every detail of that Monday morning, and my mind flashes back to that event at the strangest times. Maybe that's my way of coping with tragic or violent events — repeating them till they lose their power? Who's to say that's the wrong or right way to react? TV? Like I'm gonna take its opinion on anything!

# 28

***Coming up, Wayne Robson, Elliott Gould and Michael Murphy are featured in a special week-long tribute to the character actor on A&E's Biography ....***

There are big stars. They get big salaries to act in big movies and TV shows that suit their big egos.

The other 99 percent of actors take just about every project that comes their way — TV movies, straight-to-video releases, sitcoms, radio dramas, cartoon voice-overs, and commercials. It doesn't matter. If ya wanna eat, ya gotta work. I've grown to respect these hard-working folks. They keep plugging away, giving fine performances in good and bad projects, and rarely grabbing the spotlight. And because the TV and movie world lacks imagination, most hard-working actors get cast in roles that fit their looks or demeanor, or have similarities to a role from their past. And if you need "types" for your next project, the following might be handy ...

## The Weasel

Toronto-based actor Wayne Robson has had a lock on weasely, low life roles for more than a decade. And with good reason. He's a small, skinny guy with a high, nasal voice that he uses to great effect. He's probably best remembered as Shorty, Richard Farnsworth's dim-witted sidekick in the wonderful western *The Grey Fox* (1983). But long before that break, Robson was plying his trade in blink-and-you-miss-him supporting roles in Robert Altman films, most notably *McCabe and Mrs Miller* (1971), a personal fave of mine.

I first met Robson when he was playing weird murderer Roy Ebsary in the TV movie *Justice Denied* (1989). It was a dramatic retelling of the wrongful conviction of aboriginal Donald Marshall for a murder Ebsary committed. In puffy white wig, big geeky glasses and a Colonel Sanders beard, Robson looked very Ebsary, drawing stares from passerbys.

"So many roles I get aren't sympathetic," he said. "Either I'm the bad guy's assistant or the guy who gets blown away by the good guy. Whether it's because of my size or my shape or my looks, I'm perfectly suited to being blown away in the second act."

It thus should be no surprise that Robson got a Gemini Award for playing a weasely lowlife in the gritty TV movie *And Then You Die* (1988). Strangely, that critical attention didn't help his career immediately.

"My Gemini says on it Best Supporting Actor ... *And Then You Die*, which is what's happened to my career lately," he said with a laugh. But times change, and Robson has kept plugging away. Watch for him in reruns of Canadian shows like *The Twilight Zone* (the new one), *T and T*, *Street Legal*, *The Red Green Show*, *The Campbells*, *Chasing Rainbows* and *Scales of Justice*.

Robson was also a crazed, cackling gunman who shot Kris Kristofferson's hand in the TV comedy *Sodbusters* (1993), a drunken Nova Scotian in Robert Frank's *Candy Mountain* (1987), a bass player in the wartime drama *Bye Bye Blues* (1989), a school official in *Housekeeping* (1987), and a bumbling thief named Beano in the awful road comedy *Two If By Sea* (1995) with Sandra Bullock. And don't forget Robson as the *Bluenose* captain in one of those Canadian Heritage shorts that seem to air on TV every fifteen minutes.

His biggest role in recent memory was as Christie, a profane garbage man in the TV movie adaptation of Margaret Lawrence's *The Diviners* (1993). The difference this time was that Christie was a likable weasel. He got another Gemini for that one.

"When these kind of roles come along that are jam-packed with sympathy, it's a great change," he said. "Every actor loves these types of roles." Still, Robson was a realist. He expected to keep hustling for work for years to come.

"There's just not enough in Canada to sustain an actor. It's a hard scramble. After I did *Scales of Justice* (in 1993), I didn't work for three months. And that happens all the time. I got three weeks' work recently and that took care of Christmas."

## The Rumpled Loser

Elliott Gould was the king of rumpled cool in the late sixties and early seventies. In great flicks like *Bob and Carol and Ted and Alice* (1969), *M*A*S*H* (1970), *California Split* (1974), *The Long Goodbye* (1973), and *Little Murders* (1971), Gould perfected an uncaring, generally-well meaning slob persona. Unshaven, with a curly mop of hair and clothes that appeared slept in, Gould was the perfect antihero of the early 1970s — a nice guy who is beaten down by the system, but who really doesn't seem to care much.

Even as familiar film-noir private eye Philip Marlowe (in Robert Altman's *The Long Goodbye*), Gould remains Gould — sleepwalking his way through a murder investigation, more concerned about his missing cat than his missing friend. Being a cat owner, I identified with that Marlowe guy, and I told Gould as much.

"Well, it's nice that you like all these old things. They were fun to do," he said. "The intention was always to make a movie with Bob Altman every two years. But it didn't work out that way."

Too bad. Good roles for the actor were hard to come by after 1975, though he did find quite a few awful ones. My favorite bad Gould part is *Matilda* (1978), in which he and Robert Mitchum acted opposite a boxing kangaroo. Enough said. Still, Gould plugged away on TV and in the movies throughout the 1980s. He wasn't a leading man anymore, but he was busy. His career eventually got a boost when he played a dim lowlife in the Warren Beatty flick *Bugsy* (1991).

Just before I talked to Gould in 1997, he'd been popping up in a series of TV roles. He did the voice of a rabbi on the wonderful animated cartoon *Hey Arnold!*, narrated a nature documentary on the Komodo Dragons, and played a suitably rumpled (naturally) Albert Einstein in the Canadian TV special/pilot *The Genius*.

"I don't think I've ever before played anybody who's lived. And with Einstein, I found it tricky. You can get a little too cute and he starts looking and sounding like Geppetto."

Of course, Gould's highest profile gig in years is his recurring role as Monica and Ross' dad on the high-rated sitcom *Friends*. Not exactly the most challenging part for Gould, but a lot of people do watch that show.

"As I get older, and my characters get older and mature, I'm starting to feel there's a real future out there for me."

## The Guy Next Door

Michael Murphy tells a funny story about being on the set of *That Girl*, the goofy sixties sitcom about a single gal (Marlo Thomas) making her way in the big city.

"I just happened to be on set and Marlo comes rushing up to hug me, thinking I was Ted Bessell, her costar. And every time I watch myself in older stuff, all I can think is — yeah, I did look like Ted Bessell." And he still does. Now in his late fifties, Murphy still has the clean-cut, well-scrubbed character down pat. In his four-decade career, he's played an inordinate amount of conservative authority figures: ambassadors, police chiefs, secret service guys, politicians, and professors. He was the political advance man in *Nashville* (1975), the mayor of Gotham in *Batman Returns* (1992), and Woody Allen's two-timing friend in *Manhattan* (1979). *The Encyclopedia of Film* describes him as specializing in "angst-ridden urban types."

"I was the original whining yuppie," he concedes. "It started with *An Unmarried Woman* (1978), and then I did *Manhattan*. They were both hits, so soon that was all I was offered."

Murphy's seminal role came when he teamed with pal Robert Altman (there seems to be a theme developing here) for the 1988 political satire *Tanner*. That pay-TV miniseries, penned by *Doonsbury's* Gary Trudeau, cast Murphy as a presidential candidate on the Democratic primary trail. In its early episodes, *Tanner* mixed fact with fiction, as real Democratic candidates like Gary Hart and Pat Robertson played themselves.

"They were reluctant at first. But once they caught on to what we were up to, they wanted to be in it. So we'd just call them up." Unfortunately, the "early" episodes were all we got of *Tanner*. Though the intention was to run the fictional campaign right through to the 1988 presidential election, it didn't happen.

"It ended early. It was frustrating. Here they had Robert Altman and Gary Trudeau working together. They were crazy not to stick with it."

Murphy had no plans to try his hand at series TV. Though not opposed to the occasional guest shot (he recently played a TV movie director in the CBC series *Black Harbour*), he was reluctant to take on the weekly TV grind.

"It can be a terrible life. Remember that gal from *ER* (actress Sherry Stringfield) who walked away from the show and $70,000 a week. Everyone in Hollywood thought she was crazy. But it can be really hard doing something week after week for long hours that you're not thrilled with."

Although Murphy's TV experiences haven't always been the best, he said a recent TV movie, *The Ultimate Lie* (1996), featured one of his best performances. He played a law school dean who calls a hooker to his hotel room, only to discover it's his daughter. Though the drama had more to do with family pressures than hookers and philandering dads, it got lumped in with all the other snatched-from-the headline TV tales that year.

"And here was this good story that I was quite proud of. I thought I was very good. But then it played on TV, and no one said anything about it. It was just another TV movie. Very disappointing. But that's television for you."

## The Connection

Hello. My name is Ian. And I'm an Altmanaholic. And it's been one day since I watched something he directed ....

I hadn't intended my look at supporting actors to turn into a Robert Altman tribute. But it sort of came across that way, didn't it? Robson, Gould, and Murphy are all Altman regulars to varying degrees. Heck, Murphy can trace his association with the director back to the early sixties and the show *Combat*.

As for me, I became fascinated with Altman in my teens, when a curious movie kept appearing on late night TV. It was the grimiest, dirtiest western around, featuring windowless shacks, toothless whores, constant rain, and a Leonard Cohen soundtrack. Everyone mumbled, even the hero: a slightly dim-witted blowhard who claimed to be a gunfighter but probably wasn't. This movie kinda caught my attention when I was turning the dial at 1 A.M., let me tell you. I eventually found out that the movie was *McCabe And Mrs. Miller*, Altman's 1971 non-hit starring Warren Beatty and Julie Christie as doomed lovers in a backwater mining town. And when I finally managed to watch the whole flick, I loved it. So I checked out Altman's other work of the period — *M*A*S*H* (the movie), *Nashville*, *The Long Goodbye,* and *California Split* — and discovered a whole army of disheveled outcasts, who are good at heart, but are none too proud of it.

I could say I like Altman's realistic, overlapping dialogue, his understated humor, his fluid camera, and the improvised feel he lends to his projects. But no. That's not it at all. What drew me to Altman in the first place was his cynicism and his love of loser/heroes. McCabe is a drunken idiot, yet he saves the frontier town from the evil train monopoly. So is McCabe a hero now? Yes. Well ... sorta. I dunno. What's a hero? For that matter, what's a villain? I've never met a purely evil person in real life. Does one really exist? Of course not. People are more complicated than they're allowed to be in the movies and, most especially, on television. Forget about the flawless action hero of today. He bores me. Give me a flawed guy who fights in vain against a corrupt system, and I'm in heaven. And that's what Altman offered up — an endless array of dysfunctional heroes, outcasts who don't quite fit in, and do the right thing only for the wrong reasons. This seems more honest and more natural to me.

You can trace Altman's loser heroes all the way back to his work on the World War II TV series, *Combat* (1962-67). As a writer and director in the first season of the series, Altman presented a world in chaos, where nearly every episode began with beleaguered Sergeant Chip Saunders (Vic Morrow) nearly getting his ass blown off.

By the end of these one-hour, black-and-white war tales, someone would have died or at least grown hardened and cynical. Soldiers would betray their friends, shopkeepers their country, and five year olds would shoot officers in the back. Saunders maintained his dignity amid the chaos, but little else. He was winning the war, but at a terrible price. Everything seemed hopeless and pointless in the world of *Combat*. Heck, even the backdrop (a tangle of blown-up and burned-out buildings) suggested the war's futility. What the heck were they fighting for? This pile of rubble?

*Combat* was a great show without heroes. Even referring to Morrow and Rick Jason (who played Colonel Gil Hanley) as the "stars" of the show doesn't feel quite right. They were the main characters, but they didn't always play a main role in every episode. *Combat* was an equal opportunity employer.

Just like the war. And life.

# 29

***A sports hero from Ian's past proves less than heroic in tonight's episode of Death By Television. Acclaimed filmmaker Robert Altman directs ....***

Televised baseball has supplied me with plenty of heroes and losers over the years. But the player who best personifies the loser/hero has to be Boston-great Carl Yastrzemski.

Even before I was born, The Boston Red Sox had a history of failure. Some Beantown fans remain annoyed that their team traded Babe Ruth to the Yankees way back in 1920. But who can blame them? Nearly eight decades later, the Red Sox are still searching for a World Series championship.

Of course, it wouldn't matter all that much if the Red Sox were always inept. Your expectations would be so low, you'd never get hurt. Boston's problem is that it's a good team that fails at crucial moments. Even in a sport defined by failure, Boston stands out. The team is tragedy personified, having stood on the edge of World Series victory many times, the most recent being 1986 when Boston was one out from the championship, only to see it slip away.

Back in 1978, I wasn't aware of Boston's sad past. The Boston Red Sox of 1978 weren't as flashy as the rival New York Yankees. In fact, the Sox were kinda homely-looking, with their bushy sideburns and ill-tailored uniforms. But who wants to cheer for slick defending-champs like The Yankees? I certainly didn't. What's there to cheer for? Another championship? Where's the fun in that? I guess I had this loser/hero fetish even back then. So I attached myself to the Red Sox and, inadvertently, to the one player who best represented their tragic legacy — Carl Michael Yastrzemski.

Yaz was thirty-eight then, and well on his way to a Hall of Fame career that would see him hit more than four hundred homers and three thousand RBIs, none of which seemed to cheer him up any. I must have watched him play twenty games on TV that year, yet I can't remember him ever cracking a smile, or saying anything memorable. TV-friendly he wasn't. Heck, in the most memorable quote ever attributed to Yaz, he claimed that he "never had any fun at the game." Wow. And this guy played twenty-three seasons? Weird ...

Yet, I can't help feeling that it was Yaz's aloof personality, his lack of flash, his grim, no-nonsense approach that appealed to me back then. He had a style and non-personality all his own. He wasn't like the other over-paid and over-promoted "superstars." Yaz let his bat and glove speak for him. He was player #8. That was it. Take him or leave him. And I took him. What's not to like about a ballplayer who seems miffed that anyone likes him at all? So while my fellow students cheered the well-televised antics of Reggie Jackson and Billy Martin, I cheered for ole grumpy #8.

Maybe Yastrzemski's persona wasn't so mysterious. It might have had something to do with the fact that his personal, on-field success failed to produce a World Series Champion ring for Boston. Though bound for the Hall of Fame, Yaz still wasn't considered a "winner" in the eyes of baseball pundits. And that too appealed to me, in a Robert Altman sort of way. Who wants a guy that has everything? I want to cheer for someone with something to prove. And something to win. Or to lose.

As a paper boy, I got an early look at the baseball box scores in the summer of 1978. Before dawn, I'd hop on my bike and drive the few blocks to the newspaper drop-off. There, I'd claw open my bundle of *The London Free Press* and turn immediately to the sports section to check on my Sox.

And it looked like I'd picked the right year. The Sox came out of the gate like gangbusters. By mid-July, the Red Sox sat fourteen games ahead of the despised Yankees, who were embroiled in typical Billy Martin/George Steinbrenner/Reggie Jackson ego problems. It seemed to me like an insurmountable lead.

But I was a baseball novice, unschooled in the many ways a ball team can suddenly, inexplicably, go wrong. Injuries can play a part. So can bad hitting, bad pitching, and bad luck. But usually it's all these things and none in particular. Everyone goes into a slump at the same time for no reason. Batters can't hit, pitchers can't pitch. Scoring one run becomes difficult, two impossible.

As the season reached August, the Red Sox began losing in ways I had never thought possible. Errors, strike-outs, passed balls, and more errors. There was no end to the bad news. I'd read the morning paper, and then spend my entire paper route feeling sick to my stomach. What's worse, the whole time the Red Sox were losing, the Yankees were quietly moving up the standings. The low point came in early September when the Yankees arrived at Fenway Park four games back of the Sox. I remember these televised games very well. I'd barely settle into my easy chair and the game would be all but over. The Yankees destroyed my Red Sox, beating up on them 15-3, 13-2, 7-0 and 7-4, leaving town in first place.

But then a funny thing happened. The Boston losing streak ended. In the final two weeks, the BoSox of early in the season returned, winning twelve of fourteen games, and sending my junior high ulcer into remission. By the end of the regular season, it was a dead heat. Boston and New York had both won ninety-nine games, setting up that rarest of baseball rarities — a one game, sudden death playoff. In Boston.

Back in Alymer, school forced me to miss the first couple of innings. When the bell went, I ran home to find that the Red Sox were up 2-0 on good pitching by Mike Torrez. That lead held till the seventh, when light-hitting Yank Bucky Dent smacked a Mike Torrez pitch over the Green Monster for a three-run homer. "Holy Cow" said announcer Phil Rizutto. But the Red Sox clawed back. By the bottom of the ninth, the score favored the Yankees 5-4. With two on, two out, and Yankee fireballer Goose Gossage on the mound, Yaz came to the plate.

My hero had already hit a home run and an RBI single that afternoon. And as announcer Rizzuto reminded us earlier in the game, Yaz was a tremendous clutch hitter. But not this time. On the second pitch, Yaz swung late, popping the ball up to third baseman Craig Nettles in foul territory. The game, and my first pennant race, was over. Just like that. Yaz's status as my loser/hero remained unshaken. The man who let his bat speak for him had been let down by his bat.

Despite 452 career home runs, 3,419 hits, and a Hall of Fame induction, everyone remembers Carl Yastrzemski for that playoff pop-up. But that's baseball. Players tend to be defined by their failures, not their accomplishments. Maybe Yaz did choke in that sudden-death game. But what kind of Boston star would Yaz be if he didn't have a little bit of tragedy to go with his success? I'd certainly find him less interesting.

When asked several years ago if he would give up his Hall of Fame induction for a World Series ring, Yaz replied "probably." Perfect. Just the type of succinct, non-answer I've come to expect from ole grumpy. Man, I love that guy.

***

Here are a few more TV loser/heroes who get the crap beat out of them on a weekly basis, yet return again and again ...

**Carl Kolchak** — The intrepid, straw-hatted reporter (Darrin McGavin) of *The Night Stalker* (1974-75) never got a break. Every week, he'd get knocked on his butt trying to get the story on the latest werewolf or ancient demon attacks. Kolchak was a good, dedicated newsman, fast-talking his way into press conferences ("Uh, Mr Mayor? Sorry to change the subject, but can you comment on the recent spate of vampire-like killings?") and following leads into the city sewers. Unfortunately, sewer monsters don't give good quotes.

At the end of every *Night Stalker* episode, Kolchak would dispatch the evil being, and file his story. Editor Tony Vincenzo would then kill it, yelling at Kolchak for wasting so much time on crazy stories. That's the thanks you get for saving mankind from the supernatural. Kolchak was sort of a seventies version of *The X-Files'* Fox Mulder, but with more of a flare for fashion. Will the day come that Kolchak's white suits and straw hats are back in fashion? We should be so lucky.

**Jim Rockford** — The rumpled private dick of *The Rockford Files* (1974-80) was a cool con-artist. Rockford (played by James Garner) would lie, cheat, and bribe his way to the truth. And this approach invariably led to him getting beaten up or arrested. If he was lucky, only his car was damaged. If unlucky, he was beaten up, arrested, and his car was totaled.

**Slap Maxwell** — The sleazy sportswriter of *The Slap Maxwell Story* (1987) would start every show by getting hit in the jaw by someone Slap had smeared in a recent column. The comedy went downhill from there. However, star Dabney Coleman does excel at these type of lovable losers. Remember *Buffalo Bill* (1983-84)?

**Major Tony Nelson** — Nelson (Larry Hagman) endured great hardship on *I Dream Of Jeannie* (1965-70). When he wasn't being zapped to different time periods, he was being attacked by invisible dogs. But, it must have been cool having sexy Barbara Eden call him Master.

# 30

***It's urban nightmare time on this week's Death by Television as Johnston hides in a faceless city ...***

Do you watch more TV now that you have fifty channels? Or less? It's an interesting question. Most TV surveys show that viewers are watching less TV in the nineties than the eighties. However, those same surveys differentiate between TV-watching and video-watching. To me, they're the same thing. Why separate them?

I remember when my TV-watching went against the tide. In 1985, my TV world shrank to one channel, and I didn't mind. Really.

I'd just transferred from Ottawa's Carleton University (too cold) to London's University of Western Ontario (just right). Except for the climate, I didn't like London much. It just seemed so big, uninteresting and soulless, with too many malls and not enough good bars. Probably the best London-bashing quote I ever heard came from Pat McKenna, a hard-working Canadian comic best known as the greasy-haired nerd Harold on *The Red Green Show* and as Marty on *Traders*. McKenna had spent several months in London while working on a Second City production in London. When I asked him about it several years later, he rolled his eyes.

"I hated London," said McKenna. "My wife hated it. Even my dog hated it. It would bark at the air all the time. But the minute we moved to Toronto, it stopped."

I felt much like McKenna's dog. I lived so far off campus and so far from downtown, it seemed pointless to go out in search of fun. I tried several times in the first few weeks, then just gave up. So I stayed at home — in this case the basement of a townhouse not far from the Guy Lombardo Bridge. But basements tend not to have cable hook-ups. So for that whole year, I could only pick up one channel on my little black and white portable — CFPL, London's CBC affiliate.

I was okay with this because I grew up watching CFPL, and their supper hour newscast FYI (they had the name before *Murphy Brown*). And as regional news-casts go, CFPL was the best. It was also the only one covering our region extensively, so that's probably why. CFPL's slogan, "It's On TV London," never had so much meaning as it did in 1985. Everything I watched was indeed on TV London. By default. But after a few weeks of adjustment, I found the lack of TV choices relaxing. A critique I wrote at the time for the university paper, *The Gazette,* reflects my limited TV world, and my cynical, smart-aleckiness: "Placed up against the non-competition of a cableless television, CFPL-TV London invariably comes out on top."

Whether it was London's dullness, or CFPL's programming genius, I found myself watching more TV. Free of options, I became a loyal viewer of shows I would never normally watch. Documentary series like *The Fifth Estate*, *Market Place*, *The Nature of Things*, and *Man Alive* suddenly became must-see shows. I laughed along with Louis Del Grande's sly paranormal detective comedy *Seeing Things* (a few years ahead of its time, methinks), and tolerated the waning west coast charms of *The Beachcombers*. Late Fridays, there was that now archaic concept — the video show. *Good Rockin' Tonight* was one of the last, late night music video series to survive after Much Music's arrival.

The American imports were a real crapfest, though a watchable one. There was the violent-yet-bloodless *The A Team*, William Shatner as the aging cop in *TJ Hooker*, *Newhart*, and midnight reruns of *Dallas*. I don't like *Dallas*, but it was midnight, I was tired of studying, so what could I do? Forgive me. I was a starving TV man. JR and Sue-Ellen kept me going through the hard times.

Looking back, CFPL really didn't have a bad TV line-up. There was just enough schlock to offset the serious-minded documentaries. And when you think about it, does all that surfing up and down the dial really give you better viewing? Do a hundred channels bring more happiness than one?

Maybe not. But I think I'll stick with a hundred channels to be on the safe side ...

**Fade in ....**

**LONG SHOT** — EXTERIOR, LONDON BUS STATION. DAY
Ten-year-old Billy enters from left and, spying Johnston, runs towards him. Johnston turns and embraces the boy. Cut to ....

**MEDIUM TWO SHOT** — Ian Johnston, and little Billy.

**Billy**

Mr. Johnston! You're not leaving London? Why do you have to go? Ma likes you, and I had a lot of fun playing catch with you and stuff.

**Johnston**

I know Billy. But there are things ... things in my life that just ain't right. And until I make them right, I can't stay in one place for too long. I've got to keep moving. You understand, don't you?

**HIGH ANGLE CLOSE** — Billy as he wipes tear from eye.

**Billy**

But ... but when you make everything right, you'll come back, won't ya?

**TWO SHOT**— Ian Johnston as he embraces Billy again, holding on as he begins to speak ...

**Johnston**

You can count on it, Billy. Maybe when I return, we can go to that new mall. Or drive around for hours till we find something fun to do. Would you like that?

**Billy**

Gee. That'd be swell.

**Johnston**

Now, run along home, Billy. Your mom needs you.

[The camera begins to pull back slowly, revealing the entire station. Billy runs off, Johnston gets on waiting bus as ....]

**THEME MUSIC** begins.

**Narrator**

A fugitive reporter gets his fill of goodbyes. Of loyalties born and discarded, yet not forgotten. Of malls not visited. Where will Johnston go this time? Far away from London, for sure. But never far away from *Death By Television* ....

**Fade out...**

# 31

***Edison Carter battles the evil network-executives who would dare cancel his series in ABC's sci-fi comedy Max Headroom ....***

Quick! Name the most annoying TV character of the 1980s?

Time's up!

Chances are you came up with *Max Headroom*, that quasi-computer generated video guy who became an over-exposed product huckster. Created in 1985 as the host of a British music series, Max made it across the water a year later to sell cola and other must-have consumer goods. By the time he became a sci-fi action series lead guy — and a damn fine one too — we'd already tired of stuttering Max. It lasted only three months on ABC.

Still, it was a fun ride for British Columbia native Matt Frewer, the real face behind the computer. Unknown before Max, Frewer became the celebrity of the moment, appearing on t-shirts and magazine covers. Political cartoonists drew Ronnie Reagan as Headroom, and Max even released a Christmas song — one of the surest signs of "making it" in America.

"It was a short, sharp shock. We did twelve episodes of the ABC show, and I made the cover of *Newsweek*, And then it was over," recalled Frewer. "I never realized how big it was until I was done with it. And now everyone tells me how Headroom was this eighties icon."

Funny thing, Frewer was never typecast in the Headroom role. His career continued to flourish. He voiced the *Pink Panther* in a cartoon update, played the American political hawk Alexander Haig in the TV movie *Kissinger and Nixon*, and starred in the short-lived comedy/drama *Doctor, Doctor*. How did this eighties icon survive when so many others have faded?

"Part of the cool thing about Max was that people still thought it was a computer-generated man, not a guy in make-up. So that provided me with a certain anonymity, which was handy."

Lord praise TV viewer stupidity.

# 32

***Critics are raving about Frank's Place, a comedy with a difference from the stars and creators of WKRP ....***

TV critics can't save a show. We can write columns to the end of time declaring a drama or comedy the best thing since sliced bread. But in the end, no amount of critical acclaim can keep a TV series alive without viewer support. But we do serve a purpose. With words like "funny," "quality" "not-to-be-missed" and "feel good show of the season," critics can motivate a few people to change the channel. But it takes millions of viewers to keep a show on the air.

In the end, viewers watch because they like a show, or because its time slot suits their lifestyle. Or both. But you can forget about the role of critics as influence. We are far more useful as TV historians, reminding viewers of all those great shows from the past they never watched.

*Frank's Place* was one of those quality shows that sank in the face of critical raves. The 1987 debut drew raves for its unique, hard-to-classify portrait of a New Orleans' Creole restaurant called

Chez Louisianne. Created by Hugh Wilson (*WKRP*) and starring former Venus Flytrap guy Tim Reid, the easy-going series emphasized character over plot, featured no laugh track, and was shot with one camera and no studio audience. It was well-written, and always fascinating. You really didn't know what was going to happen from week to week.

Calling *Frank's Place* a comedy isn't quite right. Some weeks, the show dealt with heavy issues like drug use or suicide. Not exactly big laughs in that material, though the show did find humor in odd places. *Frank's Place* was eventually dubbed a "dramedy," a tough-to-classify hybrid that also included shows like *The Slap Maxwell Story* and *Hooperman*, neither of which were big successes.

I liked *Frank's Place* and I praised it in print. I didn't watch it regularly. But when I did, I remember feeling slightly disconcerted by its varying dramedy tone. However, it was fresh and original, and that takes a while to get used to on TV.

But viewers never got the chance. After starting out the schedule on Mondays at 8 P.M. (an odd choice, considering it had little appeal to younger audiences), it was moved four times during the season, confusing audiences even more. All told, *Frank's Place* lasted twenty-two episodes. Of course, we critics continue to talk about *Frank's Place* ten years down the line, dishing out guilt to all those viewers who overlooked the show so many years ago.

"TV series like that rarely end up being commercial successes," said Tim Reid, whose career is now as much behind the camera as in front. "But I think the fact that you are bringing up *Frank's Place* ten years later says a lot about the show's impact. Can you even remember the new shows that failed from last year?"

Nope. I can't.

Reid had put quite a bit of thought into his *Frank's Place* theory. What choice did he have? It's like trying to get away from his Venus Flytrap character. It ain't gonna happen. No matter what project Reid is promoting, every media interview will eventually come around to the topic of *WKRP* and *Frank's Place*. Reid even went as far as to blame the failure of *Frank's Place* on viewer brainwashing by the networks. Really.

"Brainwashing is a harsh term, but we certainly have been manipulated. We have been taught to accept mediocrity and familiarity and to resist anything different. We have allowed people with less creative minds to control our habits and taste. There's a saying that in the land of the blind, a one-eyed man is king. Well, we've got a lot of one-eyed kings in charge of things. And when someone comes in seeing with both eyes, the first thing they want him to do is cover one."

Damn. Reid was so articulate on the subject, I started feeling bad about not watching more of *Frank's Place*. What else was on TV anyway?

Oh my god. It was *Who's The Boss?* and *Alf!*

Man, we are brainwashed ...

# 33

***On tonight's I Love Lucy, Lucy sneaks into the club against Ricky's wishes. A misunderstanding leads to chaos for the roommates on Three's Company, while Hawkeye's antics anger the other doctors on the Emmy-winning* M*A*S*H.**

Syndication kills. And even the best series on earth isn't safe.

Sold around the globe to every hick station with a few bucks, classic and not-so-classic series are transformed from once-a-week "events" to daily schedule fillers, appearing three or four times a day, five days a week on various channels.

*M*A*S*H* was one the earliest casualties of syndication overkill. Considered one of the best sitcoms ever made, this eleven-year tale of the three-year Korean War has generated millions in reruns over the last twenty years. It has been alleged that not a second goes by without *M*A*S*H* airing somewhere on the globe. I believe it. Still, no series can withstand such saturation without losing something. Seen once a week, the repetitive *M*A*S*H* plots (Hawkeye's antics draw the ire of the other doctors; a patient teaches the surgeons a valuable lesson in life) aren't as noticeable. But viewed once or twice a night, you are acutely aware of just how lame, unfunny, and repetitive the show became in its final seasons.

The same can be said of *Cheers*, which has the further problem of being set in one spot (a bar). The plot never varies, making it hard to figure out why everyone watched it so much the first time. Sam Malone (Ted Danson) schemes his way into a date, Cliff makes an ass of himself, Norm makes fun of his wife, and Woody acts like the small town hick that he is. Familiarity does indeed breed contempt as far as TV shows are concerned.

"I get recognized for *Cheers* all the time," said chubby Canadian actor Derek McGrath, who had the good fortune of playing Diane's psycho paramour in a handful of episodes. It's a blip in his two-decades-plus career, but a noticeable one: "You like the money it brings. And the attention. But feel bad about it too. It's on all the time. It just gets tiresome for people. They start hating the show when they really shouldn't. And you wish they'd stop running it."

Dramas can also suffer credibility problems in syndication. The classic series *The Fugitive* featured David Janssen as a doctor on the run for a murder he never committed. It was a great series when viewed week to week. But it comes off as ridiculous on a nightly basis. It may be semi-believable that a fugitive from justice could wander into a town every week in the same ill-fitting suit and get involved in other people's problems. But viewed every night, the ludicrousness and repetitive nature of this premise becomes woefully apparent. How could a fugitive stay ahead of the cops when he's constantly getting involved in people's lives? Talk about a trail of bread crumbs.

Watching daily *Columbo* episodes, you find yourself cheering for the murderers. Sure, you still enjoy watching Peter Falk slowly unravel the murder. But when viewed daily, the plot formula wears you down. You can't help wishing Columbo will fail to find that key piece of evidence. Or that the snooty rich murderer won't incriminate himself around the rumpled detective. It's like the World War II movie analogy. You hate the Germans, yet you're tired of seeing them make the same mistakes. So you start cheering for them.

There is a syndication flip side. The shows people didn't watch the first time can now be seen in a concentrated form. Series like *Hill Street Blues*, *LA Law*, and *NYPD Blue* have complicated continuing storylines that are easier to follow when watched nightly.

And when freed from prime time competition, some shows discover new audiences. The NBC drama *Law & Order* has always been popular. But it's biggest prime time audiences didn't arrive till old reruns began airing daily on A&E.

For star Michael Moriarty, *Law & Order* reruns have brought him a celebrity he has never enjoyed before. And all after moving from New York to the decidedly non-glamorous setting of Halifax: "It's on three times a day, six days a week. Even to the people who channel surf, my face is well known." Give *Law & Order* a couple of more years on the rerun circuit, and Moriarty will need a body guard.

# 34

***Producer Gerald uses the newspapers to flush out the fugitive reporter in this week's Death By Television. But will Johnston fall for the media bait? The net begins to close on Death By Television ...***

STUDENT'S HOBBY GOES BUMP IN THE NIGHT
*The Daily News*, September 27, 1987

Late at night, if you put your ear to Ian Johnston's bedroom door, you can hear the screams of mutants, chain saws, slicing and dicing half-naked co-eds, and the groans of scientific experiments gone horribly wrong.

It's just Ian, having a little fun.

The 23-year-old King's College student is a horror aficionado who, for the past five years, has built a personal library of close to 200 creepy, knuckle-biting, hands-over-the-eyes films. For those with nerve, you can find it all among the stacks of video cassettes littering his small room: from big budget cinema releases, such as the recent remake of *The Fly*, to grade B flicks, including John Carpenter's $25,000 university film-project *Dark Star*.

While such a macabre hobby would make most parents sleep with one eye open, Ian says his family in Aylmer, Ontario finally saw things his way.

"They felt it kept me off the streets," he laughs. "They never understood it, but they accepted it. Horror is something I've always liked. When you pick up a horror movie, you never know what you're going to get. It covers the whole spectrum."

Ian can often be found in a secluded corner of metro video stores, hungry for just one more thriller to add to his growing collection. Topping his most-wanted list is the 1979 *Phantasm*. He hasn't found a copy so far, but can't wait to see a much-talked about scene where a spiked metal ball flies through an actor's head.

But even Ian has his limits. He says he's never seen worse than the *Friday the 13th* series of films, adding "what's the point of having suspense if you know the villain can't die?"

"Give me a camera," he boasts. "And I'll show you better."

Perhaps his favorite gruesome films are those made by director George A. Romero, whose morbid talents gave us the cult classic *Night Of The Living Dead*.

He's also partial to Canadian director David Cro-nenberg, whom he met while attending film class at the University of Western Ontario. One of Ian's favorite touching moments came when Cronenberg autographed an eight-by-ten glossy of a TV set with human intestines flowing out of the screen. But he had to leave it hanging on the wall at home because he didn't have any extra room when he moved to Halifax. Not after he loaded all his video tapes in the car.

"I even left my pillow," he sighs. "I just couldn't leave the tapes."

As well as full-length features films, he also has other horrible moments captured on tape, including an episode of (Watergate break-in man) G. Gordon Liddy on the game show *Password* (which Ian says is one of the worst moments in TV history), and the 1986 shuttle disaster (the explanation — "In a crisis situation, you get the best tape").

And while some might argue the artistic value of the bloody footage, Ian says you can learn a lot from movies like *Bloodbath at the House of Death*, or *Mutilator*.

"Never go into a dark house alone," he warns.

Or stay too long outside his bedroom door if your nerves are weak.

Sheesh. That newspaper profile didn't turn out so well. I sound like a blood-thirsty maniac. TV must have warped my mind. Or maybe it was just the writer. He obviously had an editorial bent (play up the gore!). *Bloodbath at the House of Death*? What the hell is that? I've never seen that flick. Does it exist? Let's check my movie book ....

> **Bloodbath In The House of Death** *(1985) — Starring Vincent Price, Pamela Stephenson, and Gareth Hunt. A horror parody in which Price battles a team of mad scientists.*

Vincent Price was in it? That's alright then...

Maybe I helped the reporter along a bit. Like anyone, I was flattered by the attention, serving up easily-manipulated quotable quotes. You've got to watch the words that come out of your mouth when you're around a reporter. Individual words look a lot different in print than when they spill out of your gob along with the hundreds of other words.

Politicians tend to defend themselves by claiming they were misquoted in local newspapers. And for some reason, the TV media embraces these "I was misquoted" stories, usually without checking with the newspaper to see if there was a misquote at all. And if there was an error with the printed quote, it's usually a matter of the politician being taken out of context, or simply saying something he didn't mean.

I'm not bitter about it though. TV loves to run print-bashing stories. And print loves to run TV-bashing stories. It's the circle of life ... Besides, I said everything that appeared in the newspaper profile. I wasn't misquoted. I was just out-of-contexted to death. But in a very nice way that made me look blood-thirsty. Which is fine. Better than boring.

And besides, the blood-thirsty article turned me into a mini-celebrity. After the story appeared in *The Daily News*, it was picked up by the Canadian Press wire. Accompanied by a photo of me — looking dark and creepy — surrounded by posters of *Day of the Dead* and *The House of Frankenstein*, the story was carried in more than a dozen Canadian newspapers.

It was my fifteen minutes. As Halloween approached, DJs across Canada would call me during drive time and get me to say something horrific. Ralph Benmergui told a national radio audience I was a "sick man," while photographers snapped my picture in graveyards. And you can imagine Dad's shock when he opened *The London Free Press* to see his Halifax-based son staring out at him, looking all creepy-like. Dad told me later his first thought was that he'd have to bail me out of jail. He was relieved to find out it was just me sounding like some psychopath in print. He was accustomed to that. And being a journalist himself, Dad encouraged me to play up my fame for all it was worth.

And I did. The article helped me worm my way into *The Daily News*, where I did some freelance work till they got tired of looking at me, hiring me full time as your humble TV writer. Apparently, *The Daily News* people weren't put off by having a closet maniac in their midst. Maybe that's a good trait for a reporter. Hmmm ....

# 35

*Here he comes that Mighty Mouse*
*Come to vanquish the foe, with a mighty blow*
*So don't be afraid anymore*
*Cause things won't be like they've been before.*
*Here I come to save the day*
*Thank goodness, Mighty Mouse is on the way*
*So let the trumpet players play*
*(ta dum dum dum)*
*For Mighty Mouse is on the way*

I choked on my corn flakes when I read the headline: "Cleric Accuses Mighty Mouse Of Drug Use." Quite a grabber, eh?

Yes, the famed Methodist minister/watchdog Donald Wildmon was at it again. The guy who condemned the movie *The Last Temptation Of Christ* sight unseen, and called for TV boycotts of everything from *NYPD Blue* to *Ellen* had suddenly turned his sights on Saturday morning cartoons. According to a June, 1988 press release from Wildmon's American Family Association in Tupelo, Mississippi, the animated superhero Mighty Mouse had been spied sniffing cocaine in an April episode.

Sound ridiculous? Well, it was. I'd seen that particular episode, and Mighty wasn't sniffing cocaine. It was just a crushed flower. And I'm not talking about a poppy. However, Mighty wasn't completely guilt-free. For while the cocaine incident never happened, there was plenty of dark, twisted stuff in *The New Adventures of Mighty Mouse* that could prove offensive to Wildmon. If he bothered to look. I hoped he didn't. I loved Mighty Mouse.

The 1987-88 season marked a turning point for Saturday morning cartoons, forcing even a hungover university student like myself (actually, it *was* myself) out of bed at 10 A.M. to catch the animated action. I first heard about Mighty from a friend in residence who raved about this twisted tale of a superhero rodent. I wasn't immediately convinced. As far as I was concerned, any show bearing the title *The New Adventures of Mighty Mouse* wasn't worth getting up early for.

I used to watch the old *Mighty Mouse* cartoons when I came home for lunch in the early seventies. With soup, a toasted cheese sandwich, and sliced bananas in milk positioned in front of me on the TV tray, I'd settle in for half an hour of cartoons on one of those regional, one-set programs for kiddies. Our host was Big Al, a paunchy, weathered, fiftyish sort in a glittery cowboy suit. He would read on-air birthday wishes, display drawings by loyal watchers, and butcher the intros to endless *Deputy Dawg*, *Bugs Bunny* and *Mighty Mouse* cartoons. You had to love this guy.

As viewed on Big Al's show, the old *Mighty Mouse* was a nicely animated cartoon, originally made for the forties' movie screens. It was a little too cornball and repetitive for me. Every episode was essentially the same — some bad guy comes along and creates havoc in Mouseville. This forces Mighty to step in at the last minute to "save the day." MM was better than some cartoons, but not in the league with Bugs or Daffy.

I soon discovered that the new version of *Mighty Mouse* was another matter. It was not only different from the old Mighty, it was different from any other cartoon on eighties TV. As the legend goes, animator Ralph Bakshi (of X-rated *Fritz The Cat* fame) convinced CBS it could save money by simply remaking an old cartoon the network already owned — *Mighty Mouse*. The network agreed, and

Bakshi set up shop, hiring a group of young animators, hungry to try something a little different from the badly-animated, mind-numbing fare that passed for Saturday morning cartoons in the eighties.

The results, as you might expect, were anything but conventional. Those tuning in to the new *Mighty Mouse* still found the lily-white rodent superhero, dedicated to keeping mice safe from everything feline-related. However, this new Mighty also liked to spend his off hours hanging out in the Safari Motel bar, where he munched his favorite food — Carp On A Stick. And the evil villains were more than just angry cats — there were TV-addicted aliens, and a fast-food franchise owner who specialized in mouse burgers. The evil feline of the original cartoon, Oil Can Harry, was transformed into a slobbering, mouse-obsessed stalker. And then there was The Cow, a masked bull who, aided by his fly henchmen, hoped to end the exploitation of the bovine community. A super villain with a conscience.

Inspired storylines included a Chipmunk parody called "Elwy and The Tree Weasels," whose hit song was a little ditty called "Twitch and Writhe." That story was narrated by Mashy The Pup, a flattened roadkill apparently unfazed by his pizza-like status. And let us not forget the *Mighty* spin-off adventures of *Bat-Bat*, a masked hero who was a bat in-and-out of costume. His loyal consort was Tick, The Bug Wonder, a shapeless, faceless, multilegged bag of blood attached to Bat-Bat's shoulder.

The jokes came fast and furious in *The New Adventures Of Mighty Mouse.* When they weren't peppering the episodes with sly in-jokes and celebrity appearances (Marlon Brando, Pee Wee Herman, Jimmy Stewart), there were sexual innuendoes. In one episode, a throwaway gag showed Mighty returning from a vacation with Minnie Mouse. Hubba hubba. And a parody of *The Honeymooners* (with Mighty and the Cow playing Ralph Kramden and Ed Norton) implied that Ed (the Cow) had been screwing Ralph's (Mighty Mouse's) wife. Yes, the new *Mighty* was plenty weird. And coupled with CBS's surreal *Pee Wee's Playhouse*, it formed one of the most subversive hours of kiddie TV since — well — ever.

And it got weirder as it went along. By the end of the second season, Mighty was battling the very medium he was a part of — Saturday morning television. In what amounted to a kiss-off episode, the rodent hero found himself trapped in a TV, bouncing from cartoon parodies of *The Flintstones* (presented as The Jetstones), to *Rocky and Bullwinkle* (with Sly Stallone as the none-too-bright Rocky, of course) to *Scooby Doo*, in which the meddling kids battled evil Mr. E — a combination of Ed Sullivan and Mr. T.

"I pity the fool that steps on my really big shooooe," Mr. E warned.

Mighty eventually broke through the TV screen, saying that "imitation is the sincerest form of television."

"Ride a bike, read a book, pester your parents," he advised, eventually admitting that "too much of one thing is bad for you, even good TV."

It ended with Mighty looking straight at the camera and confessing his dishonesty: "But enough of this lying and hypocrisy. Let's see what television is really about." Cue the commercials.

When that episode ended, I looked at my watch, and it read 10:59 A.M. Someone had just ripped the crap out of Saturday morning TV *during* Saturday morning TV. And no one noticed. Except Wildmon.

But it was more than clever and subversive writing that set *Mighty* apart. This was one of the first cartoons to hearken back to the old days of fluid, exaggerated animation. Characters' bodies altered, contracted, contorted, stretched and shrank to express a feeling, a word, or extreme pain. A beautiful woman walks into the room, and Mighty's eyes pop out of his head. Stuff like that. That form of animation is commonplace now. But ten years ago, no one was doing it on TV.

*Mighty Mouse* would eventually pave the way for the extreme adventures of a dog and a cat in *Ren and Stimpy*, a hit show that has paved the way for a rebirth in TV animation. No surprise then that one of Bakshi's hired guns for *Mighty Mouse* was the eventual creator of *Ren and Stimpy* — John Kricfalusi. The Ottawa native calls himself a self-taught animator, though he did spend a year at Oakville's now famous Sheraton College, producer of dozens of Disney animators over the years.

In the early 1980s, Kricfalusi toiled making inferior cartoons (new versions of *Heckle and Jeckle* and *The Jetsons*) before landing a gig animating Bakshi's video for the song *The Harlem Shuffle*, by The Rolling Stones.

"Bakshi is great because he's willing to give his animators a free hand, and he'll stand up for them," said Kricfalusi, whose enthusiasm for cartoons registers in his volume: he tends to yell about things he feels strongly.

Bad cartoons in the eighties usually had a team of writers *and* a separate team of animators. Kricfalusi felt *Mighty Mouse* was successful because the animators also served as writers. They understood both the cartoon genre and the narrative they could hope to put up on the screen. Kricfalusi felt that writers working separately from the animators just created problems:

"Writers might write 'Four hundred mounted Arabs come out of the forest and attack a village.' It takes a few seconds to write, but it might take a year to animate. By the time everything is sorted out, a lot of time and money has been wasted."

On *Mighty Mouse*, Kricfalusi set up old-style animation units, each working on one cartoon, and each with creative freedom. "We were very angry that first season (of *Mighty Mouse*). Animators spend years making bad cartoons. No one let us do what we wanted to do. When they did, we went crazy."

Unfortunately, CBS started wondering what the heck was going on. Kricfalusi claimed that the network watchdog group, Standards and Practices, got totally confused by the scripts they were receiving: "This stuff was beyond them. One time, they sent us a memo telling us to make sure The Cow had the right amount of teeth. How many teeth does a cow have? We had no idea." Another memo suggested that Mighty Mouse could punch other characters as long as he "punched them tastefully." That line eventually worked its way into an episode.

Despite the early hour, *Mighty Mouse* did draw some high-level critical acclaim. The show was listed as one of the Top Ten TV series by *Time* magazine. It also received an award from the lobbying group Action For Children's Television for its "pro-social values," another line that worked its way into a *Mighty* adventure. "These are the same guys (Action for Children's Television) responsible for cutting the violence out of Bugs Bunny cartoons. They made cartoons stupid."

Kricfalusi said that, in accusing Mighty Mouse of cocaine use, Wildmon totally overlooked the episode's dark finale: "At the end of that episode, when the couple is riding into the sunset, they exploded in a big mushroom cloud. Nobody noticed that."

Still, the controversy over Wildmon's accusation did cause CBS to remove three seconds from the episode in future airings. Not that there were a lot of future airings. For despite critical praise, *Mighty Mouse* petered out after two seasons due to low ratings. A line of *Mighty Mouse* toys sold by Wendy's didn't exactly catch fire, either. Heck, the hamburger chain was still trying to unload them two years after the show ended (I got a Cow action figure and one for bald master villain Petey Pate. They were all out of Mighty Mouse figures).

Kricfalusi himself didn't even make it to *Mighty's* second season. After directing eight of the twenty-six episodes in the first season, he couldn't come to a contract agreement, and just left. Still, he continues to shake up the cartoon world. In fact, after hitting even bigger success with Nickelodeon's *Ren And Stimpy*, Kricfalusi was ousted by the network in its second season after battles over scripts and missed deadlines. One story had Nickelodeon objecting to a Christmas special in which Ren and Stimpy go out in search of Stimpy's first fart. Sounds like a Yuletide family story to me.

*Mighty Mouse* rejuvenated the moribund TV cartoon industry and fooled a major network into thinking it was airing something only little kids would like. Sick little kids, maybe.

Sure, Reverend Wildmon and the network may have caught on after awhile, but it was a great ride while it lasted:

Here he comes, that Mighty Mouse
Like a bolt from the blue
With a heart that's true
Fighting evil, fighting crime
And always there in the nick of time
Here he comes to save the day
And he will prove that crime will never pay
So let the trumpet players play
(ta dum dum dum)
For Mighty Mouse is here to stay.

# 36

***Don't miss tonight's premiere of A Poke In The Eye, a video review show with a difference. And on tonight's Death By Television, chance encounters with Charlton Heston and Another World's Charles Keating bring the fugitive reporter more attention than he wants. It's a star-studded installment of Death By Television ....***

Who's the biggest star you've ever interviewed, Ian?

I get that question constantly, and I'm never sure how to answer. What's big? I mean, I get a bigger thrill out of interviewing character actors like Elliott Gould and Ned Beatty than I would for Tom Cruise. And when Sandra Bullock was reported to be in town, I had trouble getting excited. I don't like her films much, and when you don't like a Hollywood actor's work, then all you're left with is fame. And fame has little connection to talent. Or reality for that matter.

But I'd be a liar if I said I have never gone ga-ga over a celebrity. It happens. Late one Friday in 1989, the word came that Charlton Heston, Moses himself, was taking a tour of the *USS Coral Sea* in Halifax Harbour. Heston is the type of Oscar-winning actor who gets asked to make goodwill appearances on American aircraft carriers. Afterall, he is the outspoken head of the National Rifle Association. And he not only knows the United States Constitution back to front, he played the guy who wrote it. At least I think he did. If he didn't, he should have. He's played everyone else in history.

The *Coral Sea* visit came with no guarantees for an interview. As it turned out, Heston never talked to any of the press. Still, it was kinda neat to hang out with Chuck for a few moments. I mean, he's Ben Hur, the guy who starred in the first movie I liked as a kid. Not only that, he is the Omega Man, Will Penny and Major Dundee! "Take your stinkin' paws off me, you damn dirty apes!" Love him or hate him, Heston is always memorable.

As a news event, the *Coral Sea* visit wasn't. It was strictly a PR affair, aimed at bringing attention to the final voyage of the forty-two year old craft by placing a celebrity on board. A decent photo-op. Nothing more. Well-dressed in a sweater and *Coral Sea* ball cap, the aging matinee idol arrived at the ship's entrance fashionably late but in a good mood. He greeted all the ship dignitaries with a smile, looking them in the eye and making bad jokes, at which everyone laughed heartily.

You could tell Heston had done this sort of thing before. On the ship's speaker, he even hauled out an old *Ten Commandments* joke, applying it to a recent string of good weather in Nova Scotia. "I'm sure the Halifaxians have told you this is the kind of weather they usually have all the time. I've got to take credit for it. You remember that bit with the Red Sea?" The joke went over gangbusters with the bored *Coral Sea*-ites. The broad-shouldered matinee idol (I wanted to get that phrase in one more time, I don't get to use it often) then pulled another one-liner out of his bag, a self-deprecating one at that: "The admiral tells me this is the *Coral Sea*'s last cruise. It's forty-two years old. I'm a lot older than that, and I'm not retiring." OK, maybe it was his delivery on that one.

From there, The Law Giver led us inside the ship for tea with Rear Admiral Thomas Paulsen and Captain Ed Allen. While they chit-chatted about nothing, the media were led in one at a time to squat in front of the tea table and flash a few photos. I clicked away while the assembled dignitaries pretended I wasn't there.

The artificiality of these staged events bothers me, but even press conferences have a certain phoniness. The print media arrives at the appropriate time, followed by the TV news people who are nearly always late. You pick up a press kit, a short speech is given, and then the floor is opened up to questions from the media.

Which is the problem. If you really have a good question — one that will put the politician, corporate official or celebrity on the spot — you don't ask it. Why offer it up for every reporter to hear and, quite possibly, use? It's your question. You want to hold onto it. So you wait till the press conference is over, and try to get your target in a one-on-one situation. Sometimes, you can pull off this maneuver without anyone catching on. Sometimes you can't.

To be honest, news conferences are really built for TV people. Everyone waits around while they set up their cameras. After the press conference ends, the cameras are taken off their tripods and the TV people dive in for their private, hand-held, one-on-ones.

The TV people had the advantage with Heston as well. I could have written a Pulitzer Prize-winning story on Heston's visit, using all sorts of clever adjectives and metaphors. Yet, it never would have matched that TV news image of the matinee idol on the *Coral Sea* deck, looking up at the heavens (really the fighter planes) and exclaiming in awed tones "Oh my God." Wow! Just like in *Planet of the Apes* when Heston finds the Statue of Liberty. Except without the dirty words and sand-pounding.

Still, the biggest sign of fan adoration didn't come during the Heston encounter, but a visit by Charles Keating. Who is he, you say? That's what I said when the story was pitched to me. As I found out, Keating was an England-born, Connecticut-based actor who played bad-guy-turned-good-guy Carl Hutchins on the daytime soap *Another World*. I used to catch a bit of *Young and The Restless* after *Big Al* ended. But that was twenty-five years ago. So I asked around the office — is this Keating guy important? Yes. Oh yes, I was assured by a soap fan. So I agreed to meet Keating.

The site was a bar at the Halifax International Airport. Keating was on a Maritimes tour that had him appearing in several cities in the same day, so leaving the airport was not an option. It started slowly. A couple of people working at the bar came over and politely asked for an autograph from Keating, whose rugged middle-aged good looks and pony tail have made him all the rage on the soap publicity circuit. As word spread among the airport shift workers, the autograph seekers increased. A beleaguered publicist tried to form a barrier between Keating and the masses, but to no avail. Carl Hutchins was in the bar.

"We have a very strong and passionate audience in Canada," said Keating. "I went to one spot in Saskatchewan where *Another World* was the highest rated show on day-time or night-time. I'm always running into people more familiar with the details of the plots than I am."

Keating encourages autograph seekers. They're his life blood. His character's continued existence on *Another World* is predicated on how many fans write letters to him and mob him at airports: "The producers of *Another World* have the right to open every piece of mail that comes to me. And they analyze every bit of it, and are guided by it. I don't think American and Canadian consumers know the power of a postcard. The last thing the producers want to do is piss off viewers."

As the mob grew, the fans became more bold. Cocktail napkins, ten dollar bills, receipts, and parking tickets were thrust Keating's way in search of a signature. Some of the autograph seekers weren't even sure who Keating was. They'd been drawn by the growing mob. I tried to ignore the chaos, asking why a Tony-nominated actor who was trained in Shakespeare would continue to work in soaps? I assumed money was one reason (I'm not an idiot). Fame another. But was there any other reason?

"You can't be snobbish in this business. Theater is a challenge. And so is memorizing thirty pages of dialogue in a night for a soap. Soaps can be just as demanding as *Macbeth*, and I like doing both. When I first went back to soaps, some of my theater friends thought it was an absurd thing to do. But I just look at it as another challenge."

But for some actors, hanging out in an airport can be an ordeal. I was at the Los Angeles Airport waiting for a flight to Toronto one day. I was half snoozing, not really looking around. But after a while, I became aware of people walking past me to get to a guy doing a crossword puzzle. They'd just walk up, hand him a sheet of paper, he'd sign it, and they'd walk off. I took a side glance at the guy to see what was what, and there sat Jamie Farr. Wow. Talk about celebrity cheese, I thought.

He looked a lot older than his *M*A*S*H* days and, truth be told, a little bit annoyed that he had been spied. I guess his usual routine for staying incognito was to sit with his back to the crowd, and do a crossword. It hadn't worked this time. Except on me. I just left him alone. What would I do with a Jamie Farr autograph? And what am I going to say to him? Ask him *M*A*S*H* trivia?

***

My personal experiences with fame are ... well ... curious. As a print columnist (with my picture in the paper) I do get recognized. Though not as much as TV journalists, who are actually asked to sign autographs. Unfortunately, the people who always recognize me seem to be of a similar ilk — twentyish, chubby males. It's always the same story. I go downtown Halifax on a Friday night, and meet a few friends. And towards 2 A.M., bang, I run into a couple of chubby male fans. Which is OK. We need readers. But I've often wondered — what is it in my writing that attracts that audience? I thought I was gender neutral.

My best moment of fame came right out of a bad movie. I was in my favorite bar on a Thursday night after slo-pitch, when a glass of wine arrived in front of me. "That couple in the corner wanted you to have it," said the bartender. Neat. I'd always wanted someone to buy me a drink at a bar. But it wasn't as good as it sounded. The couple were nice, but it was the chubby guy who bought me the drink. I could tell that his beautiful girlfriend didn't know who the hell I was. Same old story.

***

I also got another tiny bit of fame through TV itself. Yes, I had my own TV show once. Mine. Mine. Mine. Sorta. It was a community channel, which is TV, except with teenage volunteers and bad hair.

The show began life as *Mondo Cinema*. But after seven episodes, it moved to a better community channel (more money and equipment) and was retitled *A Poke In The Eye*. It's a name that really catches your attention in the TV listings. *Poke* was a video review show in which a friend and I reviewed bizarre vids (*Rabid*

*Grannies*, *Henry: Portrait Of A Serial Killer*) that no TV critics bother to review. Still don't, actually. Stuck up bastards.

On *Poke*, we also interviewed people in the film community, and did some video-related sketch comedy. It was crude, profane, amateurish, and definitely strange, with features on everything from silly World War II propaganda films to the rise, fall, and rise of disco movies. If I had to sum it up, I'd say it was *Siskel and Ebert* with the production values of *Hoolihan and Big Chuck*. Our opening credit sequence was illustrated with a computer graphic of a finger poking into a blinking eye that spurts blood. Best part of the show. Hands down. I was also fond of our closing credits in which Godzilla cavorted to The Carpenters' song "Top Of The World." Very moving.

*A Poke In The Eye* lasted a mere five episodes before we got our walking papers. Reasons for the sudden cancellation were sketchy. One possible reason was that the show took too much time to make. Another may have been that our subject matter was a little "rough" for local cable. I don't think it was an accident that the cancellation followed an episode featuring me first in a bathtub and then tied to a bed. An unaired Christmas show saw me dressed as a drunk Santa vomiting in the middle of the road. If you want to do this on home videos, kids, use milk. It looks vomit-like on camera, and really shows up against black asphalt. I thought it was funny. With some music and clever editing, it could have been comedy gold. Or maybe not.

Still, despite its brief life nine years ago, people remember *A Poke In The Eye*. That's the power of TV. The image is everything. Of course, it helped that the community cable channel is located between CBC and CTV on the dial. So when *A Poke In The Eye* aired at six, viewers tended to stumble on our show while searching for the supper-hour news. But instead of fires, death, and political infighting, they got our documentary feature, *Porno Queens Gone Straight*.

Our real obstacle to big ratings and lasting fame was our lead-in show. I seriously doubt any of the viewers of *Blind Senior Candlepin Bowling* stuck around to watch our nasty little video show.

# 37

***While hiding out at a film festival, burned-out fugitive Johnston considers giving up. Will he turn himself in and face his Death By Television? Tune in and see ...***
***Death By Television will be followed by Twin Peaks, the critically acclaimed TV debut of film-maker David Lynch. You won't want to miss this twisted look at small town life that The Daily News calls "TV as art."***

Don't believe me simply because I'm a TV critic. My opinion no longer represents the common man. Or reality. Or the common man's reality.

I came to this realization while listening to critic David Gilmour, the former host of the slightly snooty Newsworld program *On The Arts*. He came to Halifax a couple of times to speak about movies at the city's annual Atlantic Film Festival. I don't usually go to seminars (too much like school) but I checked out Gilmour's talks. I was curious about this elder statesman of critique. Is this what I'll end up like? Is it so bad?

Gilmour proved to be burned-out and pretty up front about it. In one talk, he told a funny story about being asked to join a critics' society where the opinionated-and-paid-for-it could get together and chat. Gilmour thought this was absurd, calling the field of criticism a "silly" business that didn't warrant a society. Who wants to get together and talk about each other's reviews? Sounds awful.

In another talk, Gilmour told the festival crowd he was having trouble discerning good movies from bad: "I'm getting so jaded by movies, I'm becoming perverse. You find yourself liking bad or weird movies just because they're different." Which resounded loudly with me. Hey! That's exactly how I feel. Nice to know I'm not alone ....

But who can blame us? So much of TV is formulaic: bad, laughless sitcoms, or well-meaning, predictable dramas. It all adds up to boredom. So you start seeking out eccentricity. Bring on the freak show; I'm falling asleep. It's gotten to the point that anything a little bit different — no matter how awful — can get my heart racing. It doesn't have to be good, just weird. That's what I like. This is not an entirely terrible state for a critic to be in. You need someone out there lending a voice to the strange. I just have to be careful and keep the bizarre-me in check. If he gets loose, there's no telling what will happen.

Still, Gilmour and I are certainly not the only freaky critics. Look how many of us were fooled by *Twin Peaks*, Mark Frost and David Lynch's gone-and-somewhat forgotten bit of bizarro TV from 1990-91. It offered a stylish take on small town, west coast life, with a murder at its center. Yet it was also off-kilter, unsettling, and did I mention off-kilter? Ancient mysticism reigned supreme in the town of Twin Peaks. Everyone had a secret. People talked to logs, and were obsessed by drapes. There were giants, backwards-talking dwarves, coffee and donut obsessions, and an acting style that a friend who worked at The Stratford Festival found laughably excessive.

Even *Twin Peaks*' opening credit sequence was unlike anything seen before on TV. As Angelo Badalamenti's haunting theme played in the background, shots of cute birds shared the screen with images of logs being sliced up in the saw mill. There was beauty, violence, and utter evil in that opening. The actual show simply expanded on that theme. Yes, *Twin Peaks* was certainly, uh, something. What exactly, I wasn't sure. But that wasn't going to stop me from heaping praise on it. I was gleeful to see such weirdness on TV. A series whose central plot concerns the brutal rape and murder of a prom queen and hooker who enjoyed kinky sex, and whose fast life of sex, drugs, and otherworldly-dabbling had infected just about every male in this small town? Just who at ABC green-lighted this stylish filth? How could a jaded critic not get turned on by a series that defied TV conventions and screwed with viewers' heads at every turn?

But being weird isn't enough in the end. As *Twin Peaks* wore on, I found something lacking — a point. We all assumed there was a point, that Laura Palmer's murder would one day make sense, that a murderer with a clear motive would emerge. Even Bob, a long-haired freaky evil spirit who menaced the townsfolk, would have been an acceptable answer had there been a satisfactory explanation as to who Bob was. I'll even accept "evil presence" as long as you give me some "evil presence" rules.

But after the first year's eight episodes (including the season-ending cliffhanger) Laura's murderer remained a mystery. The following fall-season premiere gave us Bob as the murderer, though the who, what, where, and why were nowhere in sight. By early November, ratings were plummeting as audiences became frustrated and bored. Laura's dad Leland Palmer was offered up as another possible murderer. But why would Leland kill his daughter?

Because he was possessed by Bob. But who *is* Bob, fer gosh sakes? Lynch and Frost didn't have a clue. And that's the way they liked it. By the end of season two, even the most devoted fans didn't care. The *Twin Peaks* fad had petered out due to a lack of clarity. Even my growing fetish for weird for weird's sake had been tested to the max. Give me a point, or give it a rest.

***

Frost and Lynch did give it a rest. For a year. Their next TV project was even stranger and less popular than *Twin Peaks*. Called *On The Air*, the 1992 half-hour comedy was an incomprehensible tale of TV's early days, combining slapstick, one-joke plotlines and hammy acting, all together in an unwatchable stew. Or should that be an inedible stew?

Was *On The Air supposed* to be stupid? Or was it just stupid without even trying? I made it through two episodes, then tossed in the towel. I'd been through *Twin Peaks.* I didn't need this again. Once more, the audience was way ahead of me. Only four episodes of *On The Air* made it on the air.

# 38

***Coming up, Gilligan tries to buoy the castaways' spirits with his one-man rendition of West Side Story. Just sit right back and enjoy a musical laugh-a-thon on Gilligan's Island ...***

One guy who needed *Twin Peaks* was Russ Tamblyn, an Oscar-nominated actor who had been out of the mainstream so long, he had become a trivia head-scratcher:

> Question: Whatever happened to Riff of *West Side Story*?
> Answer: That's him as Dr. Jacoby, the *Twin Peaks* psychiatrist with the sixties fetish and Hawaiian-styled apartment.

Yes! Tamblyn — now chunkier and bearded — had suddenly became mainstream again!

"It's been thirty years of mainly 'Russ Tamblyn of *West Side Story* fame,'" recalled the actor. "To have *Twin Peaks* come along and be so huge is phenomenal. It's like being reborn or being given another life. Someone came up to me in an airport and said they really enjoyed my character on *Twin Peaks*, and had I ever done anything before this? I said yes. Thank you for saying that. I can't tell you what that means. It's like someone discovering you."

It's a happy story of Hollywood survival. Tamblyn began life as a Hollywood child actor, working constantly through the forties and fifties till he found fame in movies like *Peyton Place*, *The Haunting* and *West Side Story*. Tamblyn was born to the movie business in 1934. His dad was small time actor Eddie Tamblyn, who worked as an extra, and had speaking roles in flicks like *Flying Down To Rio*.

"I sort of followed my father. I did a couple of radio shows. I started working very young when I was nine or ten. The first professional thing I did out here was this play called *The Stone Jungle*, which was directed by Lloyd Bridges. I remember a talent scout from Paramount came and also Joseph Losey, the man who directed *The Boy With The Green Hair*."

That 1948 film got Rusty going in the business. Over the years, he literally grew up on the silver screen, playing Liz Taylor's younger brother in *Father of The Bride* (1950), a fourteen-year-old delinquent in the noir classic *Gun Crazy* (1949), and a vengeful soldier in the Korean War epic *Retreat Hell!* (1952). Signed to the musical *Seven Brides For Seven Brothers* (1954), Tamblyn was called upon to use his tumbling skills (honed while serving as a cheerleader at North Hollywood High) in a couple of dance numbers. Suddenly, Tamblyn was a dancer, showing off his athletic dancing skills in *Hit The Deck* (1955) and the phenomenally successful West *Side Story* (1961).

But after *West Side*, the sixties weren't kind to Tamblyn, career wise. He had a few good roles (1963's *The Haunting*, for one). Mostly though, the sixties marked career death for the actor. Things got so bad, Tamblyn's most successful late-sixties flick was probably the Japanese, Godzilla knock-off *War of The Gargantuas* (1966), in which two guys in gooey monster suits tumble around on a miniature city set.

So what happened? Was it blacklisting? Bad career choices? Bad luck? Drugs? Tamblyn insisted he just lost interest in the acting game:

"I gave up my career to get into the fine arts. I turned down parts. I was offered the role of Gilligan in that TV show. They offered me incredible contracts. I could have been a millionaire. And I turned it down. I was living in a big house in the Palisades, and I met some artists and decided I wanted to be an artist. I gave up my career. It wasn't that my career was going down hill. I just went in a different direction. I ended up doing a lot of art work and having my work at the Los Angeles County Museum of Art, Los Angeles Institute of Contemporary Art, and other places. I made it a major part of my life."

When he did work, it tended to be in forgettable fare, or in perverse biker flicks like the x-rated *Satan's Sadists*. Those for-the-money quickies turned out to be another career mistake for Tamblyn.

"I don't have a lot of memories of them. I just went in and did them. I think I saw *Satan's Sadists*. I don't mind that one. But that film actually hurt my career. It hurt me a lot in getting into the mainstream. By mainstream I mean like episodic TV or big movies. I didn't do anything mainstream for twenty years."

While making the 1982 Neil Young-directed nuke musical *Human Highway* (featuring Devo, tap-dancing, and the destruction of the earth!), Tamblyn and longtime friend Dean Stockwell moved in together, renewing their lifelong friendship. That closeness led to *Twin Peaks*:

"It came about through Dean. He invited me to sort of a birthday party/cast party for David Lynch at Dennis Hopper's house. And so I went there and Dean introduced me to David. There was a point where I said, gee, I'd like to work with you. And he said sure, next project. Things like that are said so much around here. But I told Dean about it, and Dean said David means it. Who would have known it would take three years till the next project?"

(Hey! Next time there's a party at Hopper's house, make sure you invite me.)

During the heyday of *Twin Peaks*, *Newsweek* ran a story saying Lynch carried a severed ear in his pocket during filming. Could Tamblyn verify the story?

"I read that too. I don't know if it's true. It could be. I never asked to see his ear. I've heard he's had a great fascination with death all his life. It doesn't surprise me or shock me, because I've been into fine arts so long. I've done lots of work with death and the dark side. It's just another side of life. Lynch is a quiet fellow. If you met him and chatted with him awhile, you'd think you were talking to a boy scout leader and not this deep, demented film maker."

What does Tamblyn have in his pocket now?

"Oh (laughs), I don't know. Keys and some string."

# 39

***Fugitive reporter Ian Johnston goes undercover on a movie set in a behind-the-scenes installment of Death By Television. Later, a cop becomes obsessed with capturing a violent pedophile on CBC's searing new miniseries Major Crime. Law & Order's Michael Moriarty stars ...***

I'm not a scumbag, but I played one on TV.

That's not just a clever opening line. I did play a scumbag. And it was on television. And I was damn convincing. I know what you're thinking. All TV writers secretly want to work on TV. They're just failed writers and actors who take out their career frustrations by attacking television. OK. Sure. I wouldn't mind writing for TV. I want TV to be better. What better way to fulfill that dream than to write for it myself? Cuts out the middle man.

But I never had any aspirations as a TV actor, until a good-looking woman asked me real nice. I was in the Seahorse Tavern, one of Halifax's oldest, least pretentious bars. It has loud music (usually Led Zep 4 and AC/DC's Back In Black), low ceilings, cigarette smoke, and forty-plus years of beer drinkers oozing out of the walls, benches, and sticky floor. Nice place.

As the story goes, I was standing at the bar, contemplating the pros and cons of another beer on a weeknight. Suddenly, a young, dark-haired woman in tights started chatting with me. There's always that moment when a strange woman begins talking to you before the meaning of her words register. In that split second, your mind free-wheels through a complicated series of sexual fantasies. Unfortunately her words were far less interesting. She was a casting person, trawling the bar looking for extras to appear the next day in a courtroom scene for the CBC crime drama *Major Crime*.

I was aware of the miniseries, having spent a day at a racetrack watching a film crew shoot a scene where a detective (Michael Moriarty) roughs up an alleged pedophile (*Traders*' David Cubitt). When you're a reporter, being on a movie or TV show set is boring at its best moments. There's no glamour. Just hours of sitting, waiting and avoiding the jumble of electrical cords on the floor.

Eventually, you get your five-minute interview with an actor or director who doesn't have time for you. You're the media. You're unnecessary during filming. It's only when the movie or TV show is done that you represent any importance: free advertising.

Being an extra held little additional appeal. Who wants to sit around for a day just to be a part of a crowd scene for a second of film? I've got a big ego, but not that big. I've got a job, thank you. Not a big job, but a job nonetheless. So what changed my mind? As any actor might say, I was won over by a good role: a member of the criminal line-up in a crucial court scene. Cool. Sorta like *The Usual Suspect*s and every episode of *Hill Street Blues* and *NYPD Blue*.

I agreed on the spot. It was only after another beer that I realized the young woman had insulted me in a backhanded way. Did I really look so scuzzy that I'd pass for a low-life in a criminal line-up? What do my fellow line-upees look like? I figure it was my leather coat that got me the part. I'm a smallish (five foot, eight), non-threatening guy, but in the large, floor-length leather coat, I look like I'm up to no good ...

***

## Diary Of A TV Scumbag, Day 1.

The production is housed in an abandoned Volvo plant converted into several "sets." I can still feel the grease in the air. This is a place where it's best to avoid touching the walls.

I arrive at 9 A.M., appropriately hungover. Being scuzzy and all helps me stay in character, I reason. My fellow line-upees (line-uppers?) arrive soon after. They are all Seahorse refugees, a seedy-looking lot who, like me, were enticed by the young casting woman. There are a couple of musicians, a security specialist, even a real actor. I try to convince myself I don't look nearly as frightening as these other characters. I'm sure they're thinking the same thing about me.

After getting a haircut (the koala bear tufts of hair sticking out the side of my head aren't camera-friendly, I'm told), I join our line-up at a trailer at the back of the plant where various other extras sit, dressed as cops, courtroom officials, and spectators. Like my on-set reporter job, waiting is a big part of being an extra. That's why the money (six bucks an hour) sucks. We're not being paid to act, but to be ready to go on the set at a moment's notice. It's a matter of economics. So much money is spent every day on a TV shoot, that it's crucial that everybody be ready to go at all times. It's safer to get the extras there in the morning and keep them there. If that means sitting around for some time (say eight hours), so be it. Extras don't cost that much.

By hour three, I'm wishing I'd brought a book. A big novel. Anything. Just something to pass the time in this chilly trailer. I find a dog-eared copy of *Men's Health* magazine and read it cover-to-cover. It isn't enough. The real actor does tell some interesting stories of making *Titanic* (he makes a split second appearance) but even James Cameron-is-an-asshole tales are tiring after awhile.

At 2 P.M., director Brad Turner arrives at the trailer for a look-see. He's long-haired and bearded and decidedly unlike my stereotypical portrait of a director (Which is? Bald, middle-aged Erich von Stroheim in jodhpurs, beret and monocle? Or was that Josef von Sternberg? I can never get those two straight). Turner, coming off the well-received TV movie *Peacekeeper,* is looking for a "featured" scumbag to be the guy who the victim misidentifies as a pedophile. It's a coveted role which almost guarantees screen time. Maybe even a close-up. I'm in the running because of a vague resemblance to David Cubitt's character — *Major Crime's* real pedophile. But I lose out to another leather-jacketed guy with an unshaven face, receding hairline and bullet-shaped head. He doesn't look anything like Cubitt, but he is scary.

At 3:30 P.M. we get the call. Our arrival on set brings an awed hush. The crew seem impressed by the line-up's seediness: "They don't look like actors, they look real," someone murmurs. Well, thanks. You look like crap yourself. For a few of us, the arrival on set marks the end of the line. There just isn't room in the fake courtroom for ten scumbags. Two must go. I survive the cut, though I'm asked to remove my leather coat to avoid confusion with another leather-jacketed guy. This decision is later reversed once they get a look at my puny physique. I'm relieved. My character *is* my jacket.

As Canadian TV miniseries go, *Major Crime*'s cast is strong. Besides Moriarty, the courtroom scene features Megan Follows (*Anne of Green Gables*), and an actor I always refer to as the *Hitchhiker* guy. You know the guy — he hosted that bad Canadian anthology series that runs every fifteen minutes on the specialty channels? I know he has a name, but *Hitchhiker* guy is easier.

Moriarty makes his first appearance when we're almost ready to shoot. The Detroit-born actor has to be suffering a bit of deja vu. He's just coming off six years playing a lawyer on *Law & Order*. Now he's in another fake courtroom. Though I talked to Moriarty a couple weeks before, he doesn't recognize me in my leather jacket (it's working!). He's in a good mood, though, chatting friendly-like to Lane Gates, the kid actor, and acknowledging us low lifes with a marine-sergeant routine.

"Gentleman, it's nice to see you here, all dressed up, looking suitably scuzzy," he says. We laugh. We've been cooped up for hours, so anything sounds funny.

The scene lays out simply enough. A young victim is asked by a lawyer (Megan Follows) to look at the "bad men" assembled at the front of the court and identify the man who assaulted him (Cubitt). Led through the line by detective Tallas (Moriarty), the young boy picks the wrong guy (the "featured" scumbag), who steps forward. End of scene. I play the scuzzy guy at the start of the line. It guarantees me an establishing shot, if nothing else. Maybe my mom will recognize me.

Irrational thoughts creep into my mind. I'm nervous about falling over in the middle of the shot. Or making some unnecessary gesture. So I pick a dot on the wall over the judge's bench, and focus on it. It's a technique I perfected when I was in a high school production of *Our Town*. I only had one line in that play, but I got to spend Act Three on stage, as a dead kid in a grave yard (whose appendix had burst on a scouting trip). Sitting in a chair, I was instructed not to move, even as the live characters walked about the graveyard. To help, I focused my attention at the small lamp in the lighting guy's booth at the back of the auditorium, and I got through my death scene with nary a hitch three nights in a row.

But back in the courtroom, the dot isn't doing the job. I get dizzy during a crane shot, almost losing my balance and falling over. But I survive, and the scene is repeated. After a lunch break, the scene is done again from different angles. First, the camera is placed on a dolly that's pushed up-and-down the line, capturing our faces from a low angle. The same scene is then shot from the side, the back, and the front again. We scumbags even get a close-up. I also get the thrill of having the scene "slated" two inches from my nose. That's when they slam that clapper thing to start a scene. Now there's a movie-making cliché I can tell my friends about.

For the film-makers, things aren't going as swimmingly. Behind schedule, the kid actor has almost exhausted his maximum daily work time of eight hours. Child labor laws and all. Making matters worse, an outbreak of giddiness suddenly seizes the set. During one take, the little kid actor farts, sending Cubitt into hysterics. The extras follow him, only to be admonished by the director in a very serious, we're-wasting-time-and-money, public school teacher tone. But it's too late. When kid's day is declared over, the scene remains unfinished. So we street trash are asked to return the following day for more of the same.

## Diary Of A TV Scumbag, Day 2 — More Of The Same

We arrive at ten the next morning, and hit the set at five for an hour of work. I bring a couple of newspapers, but it isn't enough. Tell me that *Titanic* story again, would ya?

## Epilogue

Airing six months later, *Major Crime* is an interesting detective drama that peters out in the second half. Unfortunately, my debut doesn't come till then, placing me in the odd position of writing a column criticizing the very part of the show that I'm in: "Maybe I've seen too many courtroom dramas, but the trial scenes felt a little like old hat, not to mention a bit unnecessary," I begin. "There are some good moments, but nothing is half as interesting as part one."

As for my own on-screen appearance, it includes a good crane-shot view of the top of my head and my white sneakers. I also end the scene, appearing in profile over Megan Follows' shoulder as the prosecution rests. It's pretty cool. My sister calls from Toronto to say she saw me. Mom is amused too. Still, I'm quite upset to discover that my fellow line-upees appear on screen with "criminal" expressions. I thought we were just supposed to stand there in line. No one told me that we were allowed to put on these heavy-lidded bad guy faces with twitchy mouths and darting eyes. By comparison, I look frozen and statue-like as I stare at my dot. I would have made a bad guy face if I'd known. I could have looked mean and memorable. I coulda been somebody.

***

For me, 1997 will always be remembered as the Year of Michael Moriarty, that curiously talented, Emmy-winning actor, best known as attorney Ben Stone on TV's *Law & Order*. In that year, we chatted four separate times about various TV and book projects.

As a teenager, my brother and I had this running argument about Moriarty. We'd seen him play a hockey goon in the effective TV movie *The Deadliest Season* (1977) and a German slowly sucked into the Nazi regime in the great miniseries *Holocaust* (1976). They were memorable, critically-acclaimed shows that benefited greatly from Moriarty's presence.

I don't know what it is about that actor that appeals to me. He has a delicate voice and a weird way of reading his lines. It's almost a deadpan delivery. Yet behind it lies deep emotion. Even when his face registers nothing, Moriarty makes you uncomfortable. Is he going nuts? Is he about to cry, or perhaps kick the crap out of someone in the cast? You can never tell. Still, it's an acting style that has its detractors. Even in the Johnston household. My brother hated the actor. Or perhaps my brother just decided to play the devil's advocate and disagree with me.

Our Moriarty Waterloo came with the airing of the TV movie *The Winds of Kitty Hawk* (1978) about the early, bold stabs at air flight. Moriarty and David Huffman played Orville and Wilbur Wright in this talky tale of the first flight. The drama was more about the brothers' troubles on the ground than in the air — sort of a psychological profile of famous inventors. Right up Moriarty's intense-troubled-guy alley. Seizing another opportunity to renew our argument, we brothers restated our Moriarty positions. Dave guffawed at the actor's every utterance. I defended him. We argued. We argued again. Heck, I despised the Kitty Hawk tale as well, but I wasn't going to let his comments go unanswered. Our debate eventually culminated in the standard insult-and-arm-punching confrontation, right there in front of the TV. I lost. I told Moriarty this story and he gave me a blank look (an intense one, as you might expect). Either I didn't tell the story well, or perhaps he was unsure how to react to a story in which a guy who hates him beats up a guy who likes him.

Moriarty is an attentive listener. But he's a far more enthusiastic talker, taking the conversation to such unexpected topics as Voltaire, Winston Churchill, eggs benedict, and U.S. Attorney General Janet Reno. Janet really gets him going, for it was Janet who put Moriarty on a new life path and brought him to Canada. As the story goes, Moriarty's career was going fine in 1993. *Law & Order* had made him a well-paid TV star and added some stability in a career that had lots of ups and downs. One day, the makers of *Law &*

*Order* decided to bring the actor along to a meeting with Reno, who was on a crusade against TV violence in all forms.

Bad idea. Moriarty was appalled by what Reno had to say. As in Canada — where federal action has been threatened if the broadcasters didn't clean up their act — Reno was considering government interference in the battle against TV violence. Moriarty took offense at Reno's position. In a backward sort of way, she was implying that what he did was making kids kill each other. Them's fightin' words.

After their talk, Moriarty blitzed the media, speaking out against what he termed Reno's "creeping McCarthyism." Moriarty being Moriarty, his personal attack was pretty emotional. Whether coincidence or not, Moriarty was out of a job a few months after the Reno meeting. He says he was fired by *Law & Order*. They claim he quit.

The unemployed Moriarty talked of a political career, but nixed that idea, and he came to Canada. He eventually settled in Halifax after making the TV movie *Calm At Sunset, Calm At Dawn*. This is where I join the story, already in progress. After hearing he was in town, I gave him a call (At the end of his answering machine message, he says "God Bless"), and when he returned my call I asked him .... well ... why Halifax?

"Two reasons, really," he replied without a hitch. "It's a legendary port. I believe in freedom, and this port was the main artery to supply the Allies with arms, food, and material to fight a dictatorship. The other is I've fallen in love with a Halifax lady. She's everything I ever expected in a woman. I can't stay away."

I was a little put off by that last comment. Though I interview people all the time, they don't usually talk that personally. Celebs are reserved around journalists. They do hundreds of interviews, even thousands. So they're expert at avoiding comments of a personal nature. Not Moriarty though. "Everything I ever expected in a woman"? Wow. I can't even imagine saying that about a woman.

When I finally met Moriarty face-to-face, he threw me off my game immediately. It was at a harness race track in Truro, Nova Scotia, where they were filming the aforementioned miniseries *Major Crime*. I had yet to do my extra stint. I was functioning only as a reporter at that time; that is, sitting off camera for three hours, watching the actors go through a half dozen variations on an arrest scene. It wasn't until the lunch break that I was ushered down to the lower seats of the track to meet Moriarty. And when the actor spotted me, he stood up and grinned broadly.

Of course, it was that Moriarty smile that served him so well in countless troubled-guy roles. Is he being friendly, crazy or non-committal? I can't tell. For the next twenty minutes, I listened as Moriarty expounded on the news of the day. He's an easy interview. It's just tough to find a usable quote amid all his observations. Actually, there's usually lots of usable quotes, just too few on the subject you're writing about.

Moriarty is, by his own admission, not the man he was in his pre-Reno days. He's spending more time enjoying life, which includes cigarettes and the downtown bar scene. He's a lot heftier than he used to be as well. And by his own admission, a lot more outspoken. Some think he's nuts or, at the very least, eccentric. When I talked to him, Moriarty viewed his change in life as an awakening, courtesy of the U.S. Attorney General.

"After I met Janet Reno, I became a warrior," he said. "A mongoose doesn't know he is a mongoose till he meets a snake for the first time. Then, with every fiber of his being, he knows it. If my opinion sometimes bothers people, well I'm sorry. Life is too short to be quiet."

Moriarty's career change appears to have fired his creative juices. He wrote a mystery novel, *The Voyeurs*, and an autobiography, *A Gift of Stern Angels*, chronicling his Reno encounter in a very loose, emotional way.

"It's a memoir of a time that changed my life. Simon and Shuster wouldn't publish it. They said it was too angry. And it is. It's a record of an artist becoming a warrior for the first time." And what a curious, busy warrior he has become. When not acting in a variety of Canada-based projects, he's made a variety of memorable TV appearances as himself. An accomplished musician, he showed up on *The Dini Petty Show* one day to do a jazz number, debated lawyer Eddie Greenspan on freedom of speech laws on Newsworld, and then declared himself an alcoholic (but a well-functioning one) on a CTV documentary.

He's a hard guy to peg. Just when he had me believing everything coming out of his mouth, he let fly with his conspiracy theory that Janet Reno was behind the cancellation of *Murder She Wrote*:

"She was obsessed with that show. She thinks that's a bad program. Violence by kids in America is not being provoked by them sitting down and watching Angela Lansbury."

Fair enough. But could a U.S. Attorney General convince CBS to cancel one of its most popular shows? Come to think of it, CBS did suddenly move *Murder She Wrote* from its traditional Sunday slot (where it had been for eleven years) to the competitive Thursday night, effectively killing it off. Hey. Maybe there's something to this *Murder She Wrote* conspiracy theory afterall. Hmmm ...

But if Moriarty has undergone some sort of conversion in recent years, it hasn't always been blissful. He spent a night in the Halifax jail after a late night run-in with the police at a downtown bar. The incident was covered extensively in the media, and he later apologized. Not long after, he was touting the virtues of the police while promoting the miniseries *Major Crime*. When he talked about the bar incident, Moriarty rolled his eyes, lamenting his new-found big-fish-in-a-little-pond status. Still, it beats being a media target in New York. "I'm glad I learned the lessons of celebrity while up in Canada."

# 40

***In his most thrilling encounter to date, Ian Johnston takes on a government body that wants to control what the public sees and hears. Can Ian negotiate the red tape and come out victorious? Will he find the one-balled man who holds the key to exonerating him? Will obsessed producer Gerald finally catch up? The only way to find out is to tune into this special, action-packed installment of Death By Television. Parental guidance is advised ...***

It was probably the word "Bastard" that started the whole messy business. Or maybe the censor board was just having a bad day. Happens to all of us. All I know for sure is that by the time the dust settled, the Maritimes looked like a repressed backward region that didn't have the courage of its own misguided convictions.

The TV drama *Bastard Out of Carolina* was a prestige project from the beginning. Based on a Dorothy Allison novel and directed by Oscar-winner Angelica Huston, the TV drama tells the unsettling coming-of-age tale of a young, poor girl (Jena Malone), and the abuse she suffers at the hands of her stepfather (Ron Eldard).

Jennifer Jason Leigh costars as the mother who overlooks the abuse in order to bring stability to her precarious existence. It was a hard-hitting subject. But not uncommon for the tube.

What was uncommon was just how good *Bastard* was at delivering the child abuse message. TV mogul Ted Turner, who financed the flick, was so disturbed by it he nixed showing it on TNT. It eventually aired on the American pay-channel Showtime.

But when *Bastard* was sent to one regional censor, AKA The Maritime Classification Board, it was rejected. This prevented it from being screened or sold in video stores in Nova Scotia, New Brunswick, and Prince Edward Island. I got wind of the story from *Bastard*'s Canadian distributor, Cinepix, who knew a publicity bonanza when they saw one. Can't say I was particularly surprised by the whole incident. Over the years, I'd had many encounters and conversations with the folks at the Maritime Classification Board.

*Bastard* joined a list of more than four hundred banned films and videos. You'll remember such Oscar-winners as *Backside To The Future*, *Ubangus On Uranus*, and *Fifty Ways To Lick Your Lover*. No wait. I'm wrong about that last one. *Fifty Ways to Lick Your Lover* was approved by the board. Good for them.

Censorship has been alive and kicking in the Maritimes almost since the birth of cinema. But Nova Scotia's chief claim to censorship notoriety came in the 1970s when the region's censors banned Bernardo Bertolucci's *Last Tango In Paris*. The case went all the way to the Supreme Court of Canada where the banning was upheld. Five years later, *Last Tango* became a regular on late night TV. What a waste of time and energy.

My personal encounters with the region's censors began in early 1989 and what was then called The Amusements Regulation Board, an arm of the Department of Consumer Affairs. I had obtained a list of the banned movies in Nova Scotia with the idea of assessing the censor's power. List in hand, I wandered through the beaded porno curtains of local video parlours in search of banned vids. I found a dozen without much effort. No surprise there. Who can keep an eye on every store's videos?

More interesting to me were the antics of the board itself. I learned immediately they don't like to be called censors because ... well ... they don't actually edit films. They rate them, or just reject them outright.

"We view the work in the same way as either you call a ball game or classify produce before it's sold to the public," said head censor Reverend Don Trivett.

Umpires and food inspectors. I liked that. However, umps and food inspectors don't work in secret like the censors. For though Trivett would talk generally about the board, he refused to divulge the names of the censors. Paid an honorarium of thirty-five bucks a half day, the board members are provincial appointees, more or less patronage positions.

"We don't give out the names," he said. "That way they can maintain their anonymity, and work with more candor." All Trivett would say was that the censors came "from all walks of life as far as that is possible." All walks, in this case, meant between the ages of thirty-five and sixty-five, with lots of free time on their hands during the day to watch porno. Hey! I could do that job. No wait. I'm not old enough. I eventually got the censors' names (they were government appointees, afterall), and my story raised a fuss for about fifteen minutes.

In general though, I've tried to leave the censor board alone. They ban the odd erotic flick, but rarely get in the way of your video watching pleasure. So why the hissy fit over *Bastard Out of Carolina*? The scene in question came near the end, when the stepdaddy beats his waifish daughter bloody, then rapes her. Very graphic, though not pornography (no insertion shots or stuff like that). It was, however, very unsettling to watch a small, helpless girl enduring this attack. But then again, it's a movie about childhood sexual abuse. Aren't we supposed to be disturbed? Kudos to Huston for creeping us out.

It turns out, the board felt the same way as I did. They were just following the letter of the law. "The board agonized over the decision for days," explained Cedric MacLeod, the board's forthright administrative officer, who was given the thankless job of fielding media calls on the controversy.

"It's a well-done movie," conceded MacLeod. "A very powerful movie. It's the type of movie that stays with you and affects you days after you see it. But it was decided that it crossed the line as to what was acceptable."

What is acceptable in the Maritimes is laid out in brief, broad guidelines: The board must prohibit any film or video containing, among other things, a scene in which "a person who is, or is intended to represent, a person under the age of sixteen years appears in a scene of explicit and exploitive sexual activity."

Which it did. *Bastard Out of Carolina* does indeed feature a sex scene with a girl under sixteen. But so did *Romeo and Juliet*. And the CBC TV miniseries *The Boys of St. Vincent*. And so did dozens of teen-sex comedies from the seventies and eighties. *Bastard* also violated the board's prohibition against films featuring "physical abuse or humiliation of human beings for the purposes of sexual gratification or as pleasing to the victim." Again, so did *The Boys Of St. Vincent*. So why was one approved and the other prohibited?

As word of the ban spread, the Maritime Classification Board became a national story, carried on just about every TV newscast across the country and into the United States. Like us, the U.S. media love censorship stories. Director Angelica Huston sent a terse comment through her publicist: "Obviously, censorship is alive and well in Nova Scotia as in the United States." She then appeared on Newsworld to point out the obvious — that it'd be stupid and dishonest to make a movie about child abuse without an actual scene of abuse.

"We wouldn't have made the movie without that scene. For me, if you're going to make a movie on this subject, you have to show this subject," she said.

The Kids Help Foundation operates a phone line for victims of abuse. Before the board rejection, the foundation had joined with Cinepix to help promote awareness of child abuse through the *Bastard* video release. Five dollars from each video was ear-marked to the Foundation. Needless to say, they were surprised by the board's actions.

"I'm no film critic, but I can assure you this film does not overstate how children are affected by abuse," said Heather Sproule, executive director of the foundation. "It's explicit and brutal, but no more brutal than what an abused child experiences."

With so much national exposure, the story just couldn't die. Cinepix lodged an appeal to the board's current governing body — The Nova Scotia Gaming Commission (Yes, I know it's weird having the folks who monitor casinos judging dirty movies, but what the heck? It's better than their old governing body — The Department of Housing). In late March, 1989, the ban was suddenly overturned, allowing *Bastard* to be released on video on the east coast. "The (Gaming) commission's seven members have reviewed it, and the appeal has been allowed," said the commission's Dennis Kerr, ending the whole incident with an overdue thud.

The reversal bothered me. There was no explanation given for how a film that quite obviously did violate the board guidelines (as written) could suddenly not violate the board's guidelines. The Gaming Commission may have been bowing to common sense, but they were also ignoring their own guidelines. The board was right in its decision, if a bit hypocritical.

So they're going to change the guidelines now? Right? Nope. Didn't happen. But that's the problem with trying to rate or censor movies. The guidelines can never be detailed enough to include a film-maker's intent. The guidelines only cover the images on the screen, not the *meaning* behind those images.

But I'm not complaining. *Bastard Out of Carolina* got released here, afterall. And that's good. And eight months after the controversy, it aired on The Movie Network to not a peep of controversy. Again, what a waste of time and energy.

***

For years, I've received a weekly list of the movies and videos rated by the Maritime Classification Board. It's fun to see what the board is reviewing and to laugh at the silly porno titles. A few years back, I was scanning the latest list when one title caught my eye — *Women Who Suck Cock and Eat Gum*. How perverse, I thought. When I was growing up, the school yard gossip told of terrible incidents of kids eating gum. As the story went, the gum would get stuck in the stomach, catching a hold of seeds. Trees then sprouted, killing the kids from the inside out. I'm not making this up. It happened to a cousin of a friend of mine. I therefore assumed that the film was about some new weird fetishist flick about women who defy the schoolyard advice and swallow their Juicy Fruit. Ewww, gross.

But it was not to be. The following week a correction arrived, typed up by some beleaguered secretary at Consumer Affairs who must have wondered why she got up that morning: "Correction:

The video title *Women Who Suck Cock and Eat Gum* should have read *Women Who Suck Cock and Eat Cum.*"

Awww crap. That's not half as interesting.

***

Boy, that censorship tale sure took a turn into the disturbing and disgusting. Let's see if I can remedy matters with something that's light and not sick ...

The goofiest censorship incident I ever covered involved the CTV affiliate ATV in Halifax, and the sci-fi series *Star Trek: Deep Space Nine.* ATV, which aired *Deep Space* at 7 P.M. on Saturdays, had been cutting violence from the show for years without anyone raising a fuss. A snip here, a snip there. Who's gonna be the wiser but the heartiest of *Star Trek* nerds? Problem was, the Fox Network in the U.S. was airing the same episode uncut at 8 P.M. So Maritime *Trek* fans with cable began noticing something was amiss.

ATV's *Trek* cuts finally came to light in November 1995, after a decision was made to edit a kiss. A kiss that wasn't just a kiss. The scene in question involved Dax (Terry Farrell), an alien called a Trill who has another creature in her belly. The gut critter is a larva type that has lived several lifetimes inside different male and female hosts. When the host gets old, the larva just moves on to a new body, bringing the memories and personalities of its past lives to its new host. The plot had Dax running into the former alien wife of an old host. As you might expect, the past feelings of the larva in Dax surface. This leads to a couple of passionate kisses.

But not on ATV. The kiss was reduced to an embrace, the scene ending before the toe-to-toe lip lock. Fans in the know were not amused.

"I know we've had complaints, but I can't say how many," said the ATV director of communications after the cut came to light. "It was edited down because of the fact it was airing at 7 P.M., in the family-viewing hour. To reduce the scene made it more palatable."

I suspect it was the word "palatable" that set a lot of people off. The following week, forty protesters assembled outside ATV's offices, waving signs protesting censorship. My fave read "Homophobia is Not Palatable." Hey! Someone's reading my articles. Cool.

“The suggestion that a kiss between two loving women is unpalatable is discriminatory and homophobic,” said protest-organizer Jane Kansas. “This censorship sends a terrible message of blanket condemnation of love and women’s sexuality.”

Fun stuff. I like playing a shit-disturbing role. However, I can’t agree with the protesters either. Turning it into a anti-gay thing was a little simplistic. Sure, ATV was getting homophobic over the sight of two actresses smoochy-wooing at 7 P.M. But was it a lesbian kiss? I mean, Dax is a Trill, not a human. She’s a female version of a species capable of taking gender-neutral aliens in their tummies. So when she kisses a woman of another species, does that count as a lesbian kiss? Or just an inter-species kiss. Or both? Or neither? I mean, she’s not kissing the host *per se*, but the thing inside her who was once her husband? So that’s OK — in a family hour sort-of-way. Isn’t it? I doubt anyone has thought as hard on this subject as me. Which is a good thing. My head hurts.

# 41

***A practical joke backfires, forcing fugitive reporter Ian Johnston to change his identity in a life-altering episode of Death By Television ....***

Several years ago, I wrote a column about how the media overused the term "guru" to describe experts. There are "media gurus," "cultural gurus," "business gurus," and "sports gurus." Gurus. Gurus. Gurus.

I always wondered how you got to be a guru. Do you have to be part of some club? Do you need to pay dues? Or does years of service in your field eventually qualify you for guru status?

If so, I couldn't wait that long. So I declared myself the "TV Guru" in one of my columns. I explained it was all a joke. I was making fun of the word "guru" by embracing it. See? I'm important and knowledgeable because I'm called TV Guru. I'm being funny. Laugh at me.

For weeks, I kept calling myself TV Guru in print, driving home the joke till the gag was played out. Trouble was, everyone forgot the joke, and kept the name. Suddenly, my friends were calling me Guru, or just “Goo” for short. Talk about a joke boomeranging. So I went with it. If you can’t beat ‘em .....

I even started running a TV-trivia column called “Stump The Guru,” encouraging people to address their inquiries to TV Guru, care of *The Daily News*. And they did. Heck, a few misguided souls even wrote to “Stump.” Now that’s going too far ...

To The TV Guru:

Under normal circumstances, I would have agreed with your ringing endorsement of the crude and rude (animated series) *South Park*.

But you should know that Global is not showing the program in its entirety. I have seen a few episodes now from the U.S. network Comedy Central.

I taped last Thursday’s (10:30 P.M.) episode off Global, thinking it would be one of the two I missed.

It was one I’d seen — about genetic engineering — but I watched it again anyway. I was surprised that, although Global saw fit to run countless disclaimers, they made such thorough and arbitrary cuts to the content.

Can someone explain to me why it is OK for the character Cartman to rant about how he’d keep his woman in line by telling her to “knit [him] a sweater or [he’ll] slap her in the face,” but it’s not OK for him to tell her to “get her bitch-ass in the kitchen and make some pie.”

Why did they censor his repeated threats to his friends that if they touched Fluffy, his pot-bellied pig, he’d “kick them in the nuts”?

I think you should let your readers know that Global is hood-winking them into thinking they are getting to see this controversial comedy when they’re are bowdlerizing it.

Sic ‘Em

Dear Sic:

First off, I compliment you on working the words “hood-winked” and “bowdlerizing” into the same letter. Good for you. I had to look them up to check their meaning.

As to why Global censored some scenes and not others, go figure. You’ve obviously spent more time thinking about it than they did. Censorship never holds up to scrutiny.

But to the subject of *South Park*-cutting — I think Global has already got your message. After several hundred viewer complaints following the premiere of the edited *South Park* (weren't five disclaimers enough, people?) Global decided to move the show from Thursday's at 10:30 P.M. to midnight Fridays, where it will air UNCUT! (The capital letters and exclamation mark are Global's, not mine).

According to Global spokesperson Shea Warrington, the network made the move not so much because of complaints, but because it couldn't sell ads during the program.

"There were no advertisers," she said. "They were nervous about the show, and the fact it was on right after *Seinfeld*," she said.

In truth, I was surprised Global aired *South Park* at all, cut or uncut. Not since *Larry Sanders* uttered the c-word twice on CBC have I been so delightfully shocked by a TV program.

And, by the way, Mr. Sic 'Em, don't cry censorship because I cut your letter down a bit. You were a little wordy.

Since that letter appeared, Global started airing daytime promos for *South Park* featuring scenes of puking, jokes about Hugh Grant's encounter with a prostitute, and poop that dances and sings. I guess networks are a little more lenient with hit shows than unproved material.

Dear Guru:

Why are commercials for local car dealers so much louder than the TV shows on Global and ASN.

Your big fan, JC

Dear JC:

I've noticed this too. It seems that all TV commercials (not just car dealership ads) are recorded and broadcast at a volume louder than the programs. On many occasions, I've turned the volume during the commercial breaks, only to increase the volume when the program returns.

A Global information officer says everything on her channel, both programs and commercials, is broadcast at the same volume. Electronic equipment specifically controls the volume, making sure everything stays at the same level. The officer says commercials just seem louder because they're designed to grab viewers' attentions.

"Commercials are a hard-sell, so they seem louder. Commercials have a sustained amount of noise, as opposed to a drama where there are pauses and periods when things are quiet," she says.

As the theory goes, if a network cuts from a soap featuring two people kissing to a Coke ad featuring pounding music and lightning-fast editing, it will sound like the volume went up. Even when it didn't.

But then again, if it seems louder, then isn't it louder? And if a TV falls in the forest, does it make a sound? And does it make a louder sound if a commercial is on when it falls?

# 42

***Fugitive reporter Ian Johnston flashes back to his days on the stage in on an emotional and nostalgic Death By Television episode. Followed later by Lexx: The Dark Zone Stories, a twisted new science fiction adventure that's not for the squeamish ....***

I'm not brave or conceited enough to be an actor. To make it in "the biz," you must be fearless, oblivious to the slings and arrows of outraged ticket buyers. You also must be an exhibitionist and an egoist to survive. Talent helps too.

Which isn't me. I've got the ego, but have to work on the fearless and talent part. Still, I took a few stabs at acting during my high school and university days. Fortunately, there's no video record of my performance in our senior class production of *The Elephant Man*. I played the elephant man's evil trainer, spouting a terrible cockney accent I'd learned phonetically. As questionable as that performance was, it beat the hell out of my second role as the bishop who helps The Elephant Man. Seventeen-year-old circus managers? Maybe. Bishops? Not bloody likely, me thinks.

I also helped write a government-financed children's play called *Canadian Crusoes*. It told the comic adventures of kids lost in the woods. For a month one summer, I played the French, comedy relief cousin (again with the bad accent), regularly making the preschool audience cry with my sad interpretation of a little boy lost. Either that, or I just frightened the hell out of them.

I remember a performance at a public picnic in Lyons, Ontario (don't look for it on a map). Then-Ontario Premier David Peterson was on hand, and took the opportunity to give us some post-performance advice on voice projection and acting. All I could think was — man, we must really suck; even the politician isn't pretending he liked our show. We probably weren't that bad. I'm probably just being overly self-conscious.

In high school, the annual talent show was the real crowd-pleaser, drawing ten times the crowd of any play I participated in. For some reason, the sights and sounds of dancing football players-dressed-as-women and repeated renditions of Bette Midler's *The Rose* are high school box office gold. Give the people what they want.

I should stop being snooty. I was in a couple of talent shows, singing and doing tired comedy sketches stolen from *Monty Python* or *SCTV*. My biggest laugh came at the talent show in grade thirteen when I poured raw eggs on my head to that *2001: A Space Odyssey* theme "Thus Spake Zarathustra." Still, pouring chicken embryos on your head to music is not acting. It's probably not even entertainment, but it beats "The Rose."

As terrifying as my own stage experience was, I still view TV and movie actors as the most fearless performers of all. You may disagree, but hear me out. No matter how good a TV or movie performance, its presentation on the big or small screen depends on dozens of other people — the director, the writer, the editor, the gaffer. The credit list is endless. The advent of special effects and technical do-dads also means actors often must perform opposite something that doesn't exist. At least not yet.

Did I say fearless? I meant crazy. Or maybe hungry. They gotta eat, afterall. Brave, crazed, hungry actors Malcolm McDowell, Rutger Hauer and Tim Curry ventured east one winter to film separate episodes of the loopy sci-fi series *Lexx: The Dark Zone Stories*. Each had been hired for his talent and international name recognition value, though not necessarily in that order. Though all three have made horrendous flicks, each has a couple of first-rate movies to give international appeal.

Malcolm McDowell is associated with *A Clockwork Orange* and perhaps *O Lucky Man*. Hauer is known as the evil replicant in *Bladerunner*. Curry has to endure life forever being known as the Sweet Transvestite from the cult musical *The Rocky Horror Picture Show*. Beyond those productions, each actor's career has been marked by good performances in good flicks, good performances in bad flicks, and bad but not boring performances in terrible flicks. Whichever way, these guys rarely go unnoticed.

*Lexx* is an ambitious, Canadian-shot series. Sold around the world, the initial four, two-hour episodes featured massive computer effects in three quarters of the footage. Whole planets and space battles were added in post-production on *Lexx*. This means the actual filming portion wasn't thrilling to look at. Just actors on bare stages.

Still, I spent an educational afternoon watching McDowell drive a prop sword into a luckless compatriot over and over again. First, there was McDowell's direct, straight-on thrust into the chest of the victim. Then the sneaky maneuver in which the victim approaches from behind, and McDowell whirls around, sword drawn, impaling the victim with his own forward momentum.

Variations went on for hours. It was fun to watch, but it really didn't look like much. Once the technical doo-dads were added months later, McDowell's sword fight took place at an altar in front of a massive, crumbling, black cathedral shrouded in mist. But for now....

"It's a bit far out there," laughed McDowell, during a break. " I know who my character is, but don't ask me about the rest of the story. I don't have a clue."

I'll try and help. McDowell plays Yottskry, a cleric who is accidentally infused with the essence of the leader of an evil empire called The Cluster. With good and evil inside him, Yottskry battles to save the planet and his soul. Or something like that. All I know for sure, is that when the effects are in place, Yottskry meets his maker by being ingested into a big brain inside the planet. Now there's something new — even for the guy who starred in *Caligula*.

"It's really kind of fun — with good and evil pulling against each other," said McDowell. "I suppose it's not particularly far away from stuff I've done in the past, but I had some free time, so why not? I'd never been to Nova Scotia."

If McDowell had it tough working with no scenery, pity Tim Curry. His one day of work is done alone, against a green screen. His image will be inserted into the action, and he will emerge as Poetman, a hologram "survivor" on a dead planet. Eventually.

"In one sense, it's a bit strange to do scenes that way," said Curry. "But in another sense, you can do them very quickly which is why I was able to do *Lexx*."

Besides convenience, Curry says it was *Lexx*'s unpredictable script that interested him in the end. "I thought it was wonderfully ambitious. So much of science fiction is dull and predictable, but this is way out there," he said. "I think it's good to encourage that kind of imagination. I'm grimly aware that *Rocky Horror* was done on a non-existent budget over three weeks, in a tiny theater. Yet it was a good idea. Sometimes it's important to show up for guys with really good ideas."

For Rutger Hauer, the role of Bog, the king of a planet of garbage offers him a rare comic turn, albeit a twisted one. As Bog, Hauer heads a society that chops off limbs and makes them into a stew with hallucinogenic qualities. It's a meaty part, so to speak. And one that Hauer dives into with relish, sporting leather and metal scraps as clothing, and a nose plug and ill-applied lipstick. His scenes — shot in Halifax and Germany — take place on the planet of Klaagya, represented on set by a small pile of garbage. Planet to be added later.

"Well, this was very unusual," said Hauer, looking less like the sleek android he was in *Bladerunner* than a chunky Gerald Depardieu. "It takes a lot of liberties — but in a good sense. This goes farther than anything I've ever done." Which is saying quite a bit. Hauer's other project in 1997 was the New Brunswick shot *Hemoglobin*, in which he played an island doctor battling inbred mutant aristocrats who live underground and feed on corpses they steal from graveyards. *Lexx* is weirder than that?

"*Lexx* is so daring and extreme .... it goes into a real funky territory. I've been given a lot of freedom with this character as well."

Actually, all three actors go to great lengths in *Lexx*. And when the special effects, sound effects, and backdrops are placed around them, darn if they don't pull off another interesting performance on spec.

***

There's an epilogue to my *Lexx* tale. Weeks after the production shut down, I met a producer at a nearly empty sound stage on the Halifax waterfront. As it turned out, there was still some touch-up filming on the go. In the middle of a football field-sized room were four guys huddled around a camera and a Volkswagen-sized brain on a table.

One guy manned the camera, another watched the monitor. The third, a puppeteer, sat on the floor below the brain, his arm jammed through a hole and poking out the top of the prop brain. On his hand was a baby cluster lizard, a Dune-like worm with sharp teeth that proceeded to chew on the brain. The fourth guy was on hand to shovel "cranial goo" over the prop in between takes to give it that realistic look.

I thought it was all very funny, and very fake-looking. I couldn't wait to see the scene on TV. What I didn't realize was that I was watching the filming of the climax of the four-part miniseries! Those four guys in the middle of an empty warehouse were concocting the thrilling ending to the fifteen million dollar series. What had been a giant, fake-looking brain in the middle of a filthy warehouse, turned out to be the nerve center for a planet-sized insect that would terrorize the Dark Zone in the finale. And here I thought I was looking at nothing. TV can be pretty amazing, eh?

# 43

***On tonight's The Waltons, John Boy discovers the shocking truth of his birth ...***

Ma and Pa Walton lived in sin? John Boy was a bastard? Just doesn't seem right. But that's what I was told in *A Walton Easter*, a 1997 TV movie that reunited America's much-loved, Depression-era family.

I watched *The Waltons* from the start — right from that wonderful 1971 Christmas special/pilot called *The Homecoming*. It was a funny, touching and realistic dramatic look at family life during the Great Depression.

I couldn't exactly understand the whole Depression thing, but I could relate to the house full of nice but occasionally troubled kids, led by aspiring writer John Boy Walton (Richard Thomas). Created by Earl Hamner, *The Waltons* was a success because it did not romanticize a desperate moment in history. People died and were financially ruined. The Waltons only hung on by the skin of their teeth, forever finding ways to keep both the lumber mill and the flatbed truck running.

*The Waltons* were a loving family, but not a perfect one. John Walton (Ralph Waite) was a cagey sort. Though the head of the family, he wasn't exactly the nicest guy around. He was capable of great anger and prejudice. Too proud by far. His dad Zeb (Will Geer) was far more lovable but, then again, he didn't have to raise the kids, did he? Olivia (Michael Learned) was a loving, if uptight mom, and Grandma Esther (Ellen Corby) was stern and religious to a fault. Why she married Zeb was hard to fathom. The Walton kids were kids. They made mistakes every week, but usually settled into bed happy: 'Night Ma. 'Night Pa. 'Night Jim Bob. 'Night Jason, 'Night John Boy. 'Night Erin. 'Night Grandpa, Grandma.

There was also a sense of the passage of time in *The Waltons* that was unique to seventies TV. The series actually progressed through the 1930s, with events like The Hindenburg crash or the outbreak of World War II affecting the family. The kids grew up, and became independent. John Boy moved away, grandpa died, and grandma suffered a stroke.

Of course, *The Waltons*' down-home, overall-clad approach did lend itself to easy parody. Every comedian from Carol Burnett to Johnny Carson took an easy shot at America's most-loved extended family. But what could you do? *The Waltons* were red hot from the beginning, knocking off the seemingly insurmountable competition of *The Flip Wilson Show*. Even I was a *Flip* fan until *The Waltons* gave me a better option. Still, the quality of *The Waltons* dipped noticeably in the later seasons. When it was canceled in 1981, most viewers had already moved on to *Dallas* or *Dynasty*.

I was re-acquainted with the series through midnight reruns on CBC in the late 1980s. Again, I fell in love with its gentle charms — till the reruns reached the sixth season or so. But even the worst of *The Waltons* rerun episodes is preferable to the current crop of TV movie reunions. For some reason, creator Hamner has allowed his Depression-era concept to skip a couple of decades in time. Which is fine, as long as the characters get older too. But in the most recent reunion, *A Walton Easter*, John Boy is newly married and he and his wife are about to have twins. Ma and Pa are celebrating their fortieth anniversary. Except this is supposedly 1969.

Do the math. The show started with the pilot TV movie *The Homecoming*. Set in 1933, John Boy was fifteen. Therefore, John Boy should be fifty-five, though he's said to be thirty-eight in the show. As well, this new time configuration would suggest that Ma and Pa didn't get married till 1929, when John Boy was fourteen. Not likely, considering the times and Ma's religious background. Early episodes of *The Waltons* said the marriage of John and Olivia in 1916, which would make it their fifty-third anniversary. What gives?

Of course, there are other time discrepancies, including the continued existence of Grandma who, in one early episode, celebrated her sixty-eighth birthday in 1934. Heck, she had a debilitating stroke sometime during World War II, yet she's still as feisty as ever. I suppose Grandma could be 103. But that doesn't explain the Baldwin Sisters, who were also old and senile back in 1933. Yet they're still around, still dotty, and still chugging back The Recipe. Maybe it's only a decade or so of script rejigging, but it bothers me. If the makers of *A Walton Easter* have no respect for the show's original concept, why should I care about the show?

I think the time has come to take matters into my own hands and put an end to this TV blasphemy once and for all time ....

## A Walton Apocalypse

It's a long, hard journey up Walton's Mountain with a body on your back. Thank the Lord it was Grandma this time.

Still, John Boy wasn't looking forward to the trip. If only he still had the ATV. But it hadn't worked in weeks. And with old Matt Foster gone or dead, there wasn't anyone left to fix the durn thing. Funny how the conveniences had fallen away in these past months. Life was now simpler — simpler than even those tough days of The Depression. No TV, no phone and, except on those rare occasions when the generator was fired up, no electricity.

Even the people were pretty much gone. The fallout had seen to that. Slowly it had seeped through the Blue Ridge Mountains and into their homes and bedrooms. A few had died right away. Others took their own way out. The Waltons just stayed the course, keeping up the old traditions in the face of a new, unfathomable reality, and always remembering that the most important thing is family. John Boy laughed out loud at that last thought. He was sounding a lot like Zeb Walton now. When had he started quoting Grandpa again? Maybe he'd been back home too long.

Grandma looked even smaller than usual lying on the flatbed, wrapped in a white sheet. Strange how such a tiny fragile woman had been one of the last to go. After her stroke everyone thought Grandma would be dead within a year. And here it was near eight decades later, and she had outlived almost all of them. Tough old bird that Esther Walton.

Of course, the same folks had thought the old flatbed truck a goner as well. But Pa's penny-pinching and Jason's mechanical skills had kept the family vehicle moving over the years. It didn't look pretty, but it'd still get John Boy and Grandma to the base of Walton's Mountain. Eventually. The first time John Boy turned the ignition key, nothing happened. He'd read a bit about trucks but, well, book learnin' ain't the same thing as getting your hands dirty. My, my. He'd done it again. Zeb's wisdom was certainly making its presence felt today. John Boy tried the key again, and the motor turned over. Second time's the charm.

Despite the grim cargo, John Boy found himself enjoying his day out. His last trip had been Mary Ellen, and that had been ... well ... he wasn't sure anymore. At least a month. Maybe more. He'd lost track. For the days were all the same. With eleven people under one roof, you get used to noise and activity at all hours. Just going to sleep used to be an ordeal: 'Night Ma. 'Night Pa. 'Night Jim Bob. 'Night Erin. There was no end.

Now the house was dead silent, save for the sound of the wind catching the screen door to keep him company. John Boy liked that thought. He pulled out his note pad, and began to write:

"The simple noises of our daily lives gave comfort to us all," he scribbled. "The sound of Ma in the kitchen preparing dinner. The distant whippoorwill calling from a crabapple tree. The morning creak of the screen door as Pa headed off to the sawmill. These were the sounds of my Virginian childhood, of growing up on Walton's Mountain ...."

Not bad. There might be a nice, little short story in there for some big city, New York magazine. They liked his down-home tales of growing up. At least, they used to. Before ....

John Boy abandoned the notebook on the seat, and put the truck in gear. It was a bumpy ride to Ike Godsey's at the best of times. Now with few vehicles and no road crews to smooth over the potholes, the back road was a washed-out nightmare. John Boy had to stop the truck on several occasions to remove dead trees from his path. The same storm that knocked the trees down was likely to blame for the collapse of what remained of Godsey's store. Looters had set fire to the place but the burned-out skeleton had continued to stand in defiance. Now it was just a pile of scorched rubble. John Boy accelerated.

The ghost caught him completely by surprise. Though he'd occasionally heard a gunshot or scream in the night, John Boy hadn't actually seen anyone beyond the immediate family in months. Yet up ahead, shambling down the side of the road was a wizened-up figure in a weathered plaid jacket and hunting cap that had long since given up its original color.

Pulling alongside, John Boy called to the ghost, but there was no response. The face was typical of the times — grayish-yellow pallor with sunken eyes — but he recognized the way the man wore his cap. It had been pulled far down on the face, the brim eliminating both the forehead and the eyebrows. The man had to tip his head back just to see.

"Yancy! Yancy Tucker!" he shouted out the window. "It's me! John Walton! Ya need a lift?" Yancy stopped, peering at the truck and its occupant for five full seconds. Then a broad, toothless smile crossed his face.

"John Boy Walton! Why, I ain't seen you in a coon's age," he burbled out of his toothless mouth, glancing back at the white sheet on the flatbed. "Uh, watcha got in the back?"

"Grandma." Yancy stared at the body for another second, taking in John Boy's words. Nodding, he smiled again.

"She was a tough old bird, eh?" Yancy cackled, slapping his knee at his own joke. "Outlived most of us. Just like she always said she would. Never liked me much though. Yup ... couldn't stand the sight of me. Need any help, ya know, with preparations?"

John Boy opened the passenger door. He doubted this sickly specter-of-a-man would be any help, but he was starved for company. Say what you want against this fast-talking horse-thief, but Yancy Tucker was a survivor. And heck, he wasn't even much of a thief. He was more of a modern day Robin Hood, stealin' from those that had more than they needed, and givin' to those that didn't have nuthin'.

Sure, there was that time Yancy got drunk on moonshine and nearly burned down the Waltons' barn. But there was also that Christmas when he brought them a turkey as a Christmas gift. Yancy said he shot it, but Grandma had seen through the ruse, and let it slide. The family was in desperate straits back then. A Christmas turkey was welcome news. Besides, Yancy was kin. A distant cousin, mind you, but kin nonetheless. And The Waltons stuck together.

The pair traveled for a few miles in silence. There was no need for reminiscing about recent times. Their stories would be much the same. Though well past his thieving heyday, Yancy was still capable of a surprise or two, removing from his filthy coat a full mason jar of clear liquid.

"What's that?" asked John Boy in amazement.

"It's The Recipe. The Baldwin Sisters gave me a case of it several Christmases ago," he lied.

The dotty Baldwin sisters had died years before under mysterious circumstances. They'd been found in their mansion parlor, dressed in their best, seemingly happy and content, but very dead. The doctor thought natural causes, but that didn't explain what had happened to their vast stores of "The Recipe." The moonshine had disappeared, and with it the much-sought-after formula.

"My god," exclaimed John Boy, recognizing that familiar odor. "I never thought I'd taste this again. Here's to the Baldwin sisters," he toasted, taking a long swig.

John Boy stopped the truck and lifted Grandma's body from the back. But even with a light load and Yancy to help, the rest of the trip up Walton's Mountain proved to be slow going. John Boy had been feeling poorly lately. And the sickly Yancy had proven to be the hindrance John Boy had expected. Still, they covered the distance to the graveyard by early afternoon.

Pa had always intended to build a retirement home up here but had never found the time. Maybe it had just been a dream, anyway, one of those things you always talk about but never quite get around to doing. Like John Boy's dream of moving back home. His writing career couldn't possibly have flourished in small town Virginia. What was he gonna do? Dust off the antique printing press in the shed and revive his *Blue Ridge Chronicle*? It hadn't happened. It wasn't going to happen.

He'd ended up back with his family accidently anyway. Despite growing world tensions, John Boy had made the trip from Washington for Ma and Pa's fiftieth anniversary. He had misgivings, but this was a big family event. Non-attendance was not an option in the Walton family.

And after *it* happened, there wasn't anything in Washington to go back to. So John Boy had stayed put, moved into his old room, and fallen back into the old routines. It was comforting in one way. But for a grown man, being a Walton, with all its Walton rules and traditions, could be darn annoying at times. Still, he'd gone along with it all, settling with ease into the family role of oldest son, offering only half-hearted protest when someone called him John Boy instead of John Walton. Old habits die hard.

Digging the grave took over an hour. Neither John Boy nor Yancy were up to the task. Every five minutes of shoveling was followed by a lengthy break. When they finally did roll Grandma's body into her final resting place, they both collapsed with exhaustion. John Boy finally came around at sunset. Getting up quickly, he tapped Yancy on the shoulder. They'd have to work fast if they wanted to fill in the grave and get back to the truck before dark.

But it wasn't going to happen. Yancy was dead, the jar of The Recipe still clasped in one dirty paw. John Boy felt no sadness. He'd been through this many times. To tell the truth, he was surprised his distant cousin had made it this far. But then again, Yancy always had a knack for survival. Until now, of course.

Taking The Recipe jar from the dead man, John swallowed a swig and sighed aloud. He was dog-tired and feeling poorly. By the time he dug another grave for Yancy, it would be dark. And the nights can be very cold on Walton's Mountain ...

But he couldn't just leave Yancy here to rot. And he definitely didn't want to come back up. Sigh. John Boy took another swig. He thought for a moment. Would it be such a crime? Grandma would understand, wouldn't she? She wouldn't want her grandson to freeze to death. And she wouldn't want him to have to make an extra trip, would she? Of course, Zeb Walton would insist it wasn't "The Walton Way" to shirk responsibilities or to do a half-hearted job. But then again, Zeb also liked to say that the first responsibility was with the family.

Grandma would say. Grandpa would say. Ma. Pa. John Boy's head throbbed from all the pondering. He had to stop looking for advice in the past. It was just him now. He made the decisions. He was the Waltons. *John Walton*, dammit. He was nobody's Boy anymore. And the time had arrived to do what was best for his family of one.

He grabbed the dead man by the arms, and flung the body into the hole and onto Ester Walton. "'Night Grandma. 'Night Yancy," John Walton muttered to no one, depositing a shovel of dirt on his kinfolk.

# 44

***Don't miss M*A*S*H, and what critics are calling "the TV event of the week" and "a three-Kleenex season ender that's not-to-be-missed."..***

My favourite TV death comes without tears and only the slightest expression of grief. That's why I like it.

At the end of *M*A*S*H*'s third season (1974-75), actor McLean Stevenson (fun-loving nebbish Colonel Henry Blake) had decided to leave the show. Stevenson wanted to move on to other things, including a new series. With Stevenson's departure imminent, *M*A*S*H*'s producers looked for an appropriate sendoff for Blake. And what they came up with was anything but conventional TV.

The plot has the Colonel receiving his orders to go stateside. After a humorous, emotional goodbye with his friends, Blake dons his civvies and hops into a helicopter and leaves Korea for Illinois and his family. The happy reunion doesn't take place. In the program's final scene, Radar announces in the operating room to the remaining surgeons that the Colonel's transport plan was shot down over the Pacific. There were no survivors. Colonel Blake is dead.

I felt like I'd been hit by a two-by-four when that episode first aired. It was sad enough to lose a much-loved character. But to have that character killed, without reaching his goal? That hurt. Many viewers complained bitterly of the sudden, unexpected death. They found it cruel and unnecessary. Like death in real life.

Most TV deaths are carried out on screen, and come with much foreshadowing or justification. There's a closure because we can see them a mile away. Sometimes they're even announced in on-air promos days in advance. *Dallas*' Bobby Ewing gets hit by a car and killed by the woman who already tried to kill him a couple of seasons past. But wait! It was all a dream. Bobby's OK. He's just having a shower. Singer Caitlin Davies (Sheena Easton) marries Sonny (Don Johnson) on *Miami Vice*, but you just know she won't be around long. You need Sonny out there circulating with a different Florida beauty every week. Heck that's why we tune in. So Caitlin must die, preferably at the end of the season during the spring Sweeps.

But Colonel Blake's death was different. There was no lead up. TV-wise, it was illogical. Why not keep the character alive for a possible return? You'd get big ratings. The death announcement scene was also unique. Since it was set in an operating room, the death notice can receive very little reaction from the doctors. They have their hands full with dying soldiers and simply can't react beyond a few furtive glances with their eyes, the only part of their faces visible above the surgical masks. Fade to black. End of season. There's no funeral for Blake. No wake. No words of support. No flashbacks. Blake's death just happens, and people move on. This seems honest to me.

When my father died suddenly of a heart attack at the age of fifty-nine, my first reaction was not sorrow but confusion. My mind raced through my life as I tried to figure out exactly where I stood with the guy. We'd had good and bad moments, of course. As Theodore Cleaver asserts in *Leave It To Beaver*, "The only fathers that don't yell at their kids are on television." But just how all those good and bad moments stacked up in the end wasn't clear to me.

About three weeks after my dad's sudden death, I was in Washington at one of those *Famous Monster* conventions. Through a series of coincidences, I suddenly found myself sitting in a bar, chatting with the spawn of Boris Karloff, Bela Lugosi, Lon Chaney, and Dwight Frye.

They'd been called in to talk about their famous relatives and their legacy. Real estate agent Sarah Karloff said she didn't care for her father's scary movies, preferring the "straighter" roles in his long career. Lawyer Bela Lugosi Jr., who looks quite a bit like his famous father, seemed slightly annoyed by questions about his father's work with no-talent director Ed Wood. Son Lugosi felt Wood exploited his dad at a vulnerable time in his career. Ron Chaney couldn't say all that much because he'd never met Lon Sr., and was only a kid when gramps Lon Chaney Jr. died. All he could remember was wrestling with his big grandpa a few months before the actor's death.

Dwight Frye Jr. was much the same. His dad was a talented character actor best known as the original Renfield (the guy who eats the flies) in *Dracula* (1931). He died in 1943, when Frye Jr. was still a kid: "It's strange — all these questions about my father. But most of my memories are those of a young boy. I don't know as much about his career as some of the fans here. I know him best as a father."

I understood where Frye was coming from. But I also gained some insight of my own. I was a kid for the whole time I lived with my father. By the time I was an adult, I'd long since moved away, only seeing Dad during holidays. So when he died, I was confused about our relationship.

Most of my memories come from the days when I was a pain-in-the-ass caught up in my own perceived troubles and daily crises. Looking back, I know now that I couldn't figure out what kind of a dad I had because I was still trying to discover what kind of a person I was. And by the time I started figuring things out, Dad was gone. Of course, there's no guarantee Dad and I would have come to any conclusion even with three more decades. Look at Dwight Frye Jr. He's still learning stuff about his Dad five decades after his death.

Besides, moments of self-revelation are a TV concept. Confusion is a natural state for kid-parent relationships. Death just brings that confusion out in the open. And like Colonel Blake's demise, death tends to hit you when you least expect it. The telegram arrives while you're elbow-deep in a soldier's belly. You feel bad, but there's little time to stop and think about it. You've got to keep going and sort out your true feelings on the fly.

# Epilogue

**LONG SHOT** — top of a desert mountain. **DAY.**

**INTO SHOT** stumbles a man. And as the figure comes **TOWARD CAMERA,** we recognize him as Ian Johnston. He continues to run out of frame. Cut to ...

**LONG SHOT** — cliff at top of mountain.
Johnston enters frame and stops at the edge of cliff. He turns around, flinching at the sound of a gunshot. He's trapped.

**CLOSE SHOT** — Producer Gerald moving **TOWARDS CAMERA**, holding a gun.

He grins, and moves forward out of frame.

**TWO SHOT** — Gerald and Johnston, who now stand six feet apart on cliff.

**Gerald**

[*angrlly, between breaths*]
This is too quick. You deserve to die slowly.

**Johnston**

For what? For writing bad things about your TV show? A sitcom about a modern-day witch who raises a house full of orphans? How lame. How derivative. How mind-numbing.

**CLOSE SHOT** — Gerald.

**Gerald**

Oh you can come up with better insults than that, Mr. Critic Man. Say what you want, but *Cauldron For Eight* has been on the air for five years. What does that tell you?

**TWO SHOT** over Gerald's shoulder, with Johnston in the background.

**Johnston**

It tells me the public are idiots. Where would you be if you hadn't nabbed Suzanne Somers as the star, Mr. Producer? Not on a network, that's for sure. You would have been lucky to sell it in South America.

**CLOSE-SHOT** — Gerald.

**Gerald**

[*angry, cocks his pistol*]

You just don't get it, do you? You were the one that started this, calling the show [pulls a yellowed review from his shirt pocket] "sophomoric piffle for those pathetic souls who still watch *Three's Company* reruns." Funny stuff.

**CLOSE-SHOT** — Johnston

**Johnston**

I don't use the word "piffle." Much. And besides, I didn't write the review. I told you that. And I told the police. And the court. Someone broke into the office and changed the review. I didn't write it.

**CLOSE SHOT** — Gerald.

**Gerald**

Yeah, yeah. The phantom One-Balled Man. Show a little back-bone, will ya? Stand by your review. Admit your crime, Mr. TV Guru!

**TWO SHOT** — Gerald and Johnston.

**Johnston**

[*cringing at mention of the name "TV Guru"*]
Oh, I'll do it. I'll do it alright. But only when you admit you're a TV hack! Do you hear me?

[Johnston lunges at Gerald. Caught off guard, Gerald falls backward, sending the gun to the ground. The two square off, exchanging punches as they fight precariously close to the cliff edge. A hard blow from Gerald, and Johnston falls to his knees.]

Cut to ...

**HIGH ANGLE — CLOSE SHOT** of Johnston.
Johnston looks up at camera as shadow of Gerald falls over him.

**LOW ANGLE – TWO SHOT** — Gerald with gun standing over a kneeling Johnston.

[Gerald pulls the trigger; the gun lets out a quiet click. No more bullets. Gerald can't believe his luck. Without hesitation, Johnston grabs Gerald in the crotch, squeezing and twisting hard till the producer falls to the ground. Suddenly, Johnston eases his grip, and stares at his opponent's face. The secret is revealed.]

**EXTREME CLOSE SHOT** — Johnston's face

**Johnston**

It's you! It was you all along. You're the man with one testicle.

**CLOSE SHOT** — Gerald, who does not reply. Cut immediately back to ...

**MEDIUM SHOT** — Johnston as he stands up.

**Johnston**

You! You changed my review? You ... you rewrote my words to make your own TV show look bad. Why? Why?

**CLOSE SHOT** — Gerald.

**Gerald**

Because I had to. I had to sabotage the show. There was no choice. It had been on five years. It looked liked it'd be on the air forever. But I had a shitty deal with the network, and Somer's contract demands were out of sight. Hell, the whole cast and crew wanted raises. It just wasn't profitable to keep making the show. I needed to cancel it and sell the reruns in syndication. That's where the real money is. So I wrote your bad review; I had to kill the show somehow.

**CLOSE SHOT** — Johnston.

**Johnston**

Sabotage. But why did you keep coming after me? You got what you wanted — the show's been canceled.

**TWO SHOT** — shoulder and head of Johnston in foreground. Gerald in background.

**Gerald**

[*begins to rise to feet*]
You were the weak link. As long as you were around, there was still one person out there who might know the truth.

**CLOSE SHOT** — Johnston

**Johnston**

But ... why didn't you just wait for my review. I might have hated the show.

**CLOSE SHOT** — Gerald

**Gerald**

You! Ha! You had five years for that. You're burned out. You haven't hated anything in years.

**TWO SHOT** — Gerald and Johnston.

**Johnston**

But nobody listens to critics, anyway. Besides, you could have just canceled the series. It's a lot less convoluted. You could have said that you wanted to end the show on a high note. Have a big finale, get lots of publicity ....

**CLOSE SHOT** — Gerald.

**Gerald**

[*Hesitates. The idea had obviously never occurred to him.*]
Shut up!

**CAMERA** stays on the increasingly agitated, sweating Gerald as Johnston continues to talk ...

**Johnston**

You'd have been guaranteed big ratings. You could have killed some of the characters off in spectacular ways. Or maybe had a couple of them get married. That always works with the viewers. And you could have used the press coverage from the final episode to increase the show's value in syndication ...

**EXTREME CLOSE SHOT** — Gerald

**Gerald**

Shut up! Shut up! Shut Up!

**TWO SHOT** — Gerald and Johnston
[Gerald lunges at Johnston wildly. Johnston sidesteps him, and the producer keeps going, out of frame. Cut immediately to...]

**LONG SHOT — LOW ANGLE** from base of cliff looking up to top where ...

**A BODY** suddenly appears, plummeting and screaming. Before impact, cut to ...

***MEDIUM SHOT — LOW ANGLE*** of Johnston peering over cliff.

[*Johnston flinches at the sound of impact.*]

**Johnston**

[*pulls tape recorder from shirt pocket ... pushes Stop button*]: Too bad. He was a good producer. A lousy villain, but a good producer.

**HIGH ANGLE LONG SHOT** — Johnston on mountain top as he peers over cliff. Slowly walks away from edge and begins to climb down the hill. Hold on shot as theme music begins, followed quickly by voice of narrator ...

**Narrator**

May 26th. The final day of Sweeps. And the day the running stops. One man's death has brought another a new life. But can Johnston go back? Is there life after series TV, or just endless guest shots, straight-to-video releases, and "Where Are They Now?" profiles in *TV Guide*? For Ian Johnston, the future holds no promises. OK. There's a book deal in the works. But who reads books these days, anyway? This has been *Death By Television*. Good night ...

**HOLD** for the musical climax and ...

**FADE OUT ...**